T0135764

Instationary Vibrational Analysis
for Impulse-type Stimulated Structures

von Bastian Kanning

Dissertation

zur Erlangung des Grades eines Doktors der Naturwissenschaften

– Dr. rer. nat. –

Vorgelegt im Fachbereich 3 (Mathematik & Informatik)

der Universität Bremen

im August 2012

Datum des Promotionskolloquiums: 29. November 2012

Gutachter: Prof. Dr. Peter Maaß (Universität Bremen)
          Prof. Dr. Hans-Georg Stark (Hochschule Aschaffenburg)

Bibliografische Information der Deutschen Nationalbibliothek

Die Deutsche Nationalbibliothek verzeichnet diese Publikation in der
Deutschen Nationalbibliografie; detaillierte bibliografische Daten sind
im Internet über http://dnb.d-nb.de abrufbar.

ISBN 978-3-8325-3326-7

Logos Verlag Berlin GmbH
Comeniushof, Gubener Str. 47,
10243 Berlin
Tel.: +49 (0)30 42 85 10 90
Fax: +49 (0)30 42 85 10 92
INTERNET: http://www.logos-verlag.de

Abstract

Ingenieurwissenschaftliche Aspekte der Strukturdynamik sind seit je her Motivation für mathematische Forschung, und zuverlässige Prognosen über das dynamische Verhalten mechanischer Systeme stellen noch immer eine Herausforderung dar. Die vorliegende Arbeit beinhaltet sowohl eine umfassende Darstellung der grundlegenden Probleme, als auch neue mathematische Fragestellungen in der Strukturdynamik.

Nach einer Einführung in die Thematik und einer Motivation der mathematischen Modelle wird gezeigt, wie die partielle Differentialgleichung 4. Ordnung zur mathematischen Modellierung schwingender mechanischer Systeme mit finiten Elementen räumlich diskretisiert und somit in eine gewöhnliche Differentialgleichung 2. Ordnung überführt wird. Anschließend werden die Grundlagen der Lösungstheorie der entstehenden Gleichungen und ein Dämpfungsmodell vorgestellt.

Ein Überblick über inverse Probleme in der Strukturdynamik bilden den Abschluss, wobei auf 3 spezielle Fragestellungen näher eingegangen wird. Zunächst wird das sogenannte *Model Updating* zur Identifikation von Modellparametern vorgestellt. Hierzu wird ein passender Vorwärtsoperator definiert, die Fréchet-Ableitung und deren Adjungierte berechnet. Danach wird auf die Anregungserkennung eingegangen, wobei die Schlechtgestelltheit des Problems gezeigt wird und die Adjungierte des Vorwärtsoperators berechnet wird. Zusätzlich werden mögliche Maßnahmen zur Diskretisierung für die numerische Lösung besprochen. Zum Abschluss wird ein Verfahren zur Rest-Schwingungsreduktion durch Anregungsmanipulation präsentiert, das sogenannte *Input Shaping*.

Abstract

Structural dynamics constantly yields as a motivation for mathematical research, and reliable prognoses about the dynamical behavior of mechanical systems is still challenging. This thesis contains a comprehensive presentation of the fundamental problems as well as novel mathematical questions in structural dynamics.

An introduction to the theory and a motivation of the mathematical models are given in the beginning. Afterward, it is shown, how to spatially discretize the partial differential equation of 4th order that serves as a mathematical model for the dynamical behavior of mechanical systems by finite elements and transfer it to a ordinary differential equation of 2nd order. The basic solution theory for the resulting equations is presented, and an adequate damping model is introduced.

Finally, an overview of inverse problems related to structural dynamics is given, while subsequently 3 of these receive special attention. The first of these is the so-called *model updating* for identifying model parameters. An according forward operator is defined, the Fréchet-derivative and its adjoint are computed. The second discussed inverse problem is the identification of load transmissions. Here, the ill-posedness is shown and the adjoint of the forward operator is computed. Additionally, two possible steps towards a numerical solution are presented. Finally, the so-called *input shaping*, a method to reduce residual vibrations by manipulating load transmissions, is reviewed.

Danksagung

An erster Stelle möchte ich mich bei Prof. Dr. Peter Maaß dafür bedanken, dass er mir im Rahmen der Stipendiatengruppe "Scientific Computing in Engineering" (SCiE) ermöglicht hat, dieses spannende Thema zubearbeiten. Weiterhin möchte ich ihm und Dr. Iwona Piotrowska-Kurczewski für Motivation und Betreuung bei dieser Aufgabe danken. Ich bedanke mich bei Linghan Li und Dr.-Ing. Christian Schenck vom Bremer Institut für Strukturmechanik und Produktionsanlagen für den regelmäßigen Austausch und die Denkanstöße. Für finanzielle Unterstützung danke ich der Universität Bremen und der Deutschen Forschungsgemeinschaft.

Ich möchte mich bei allen Mitgliedern des Zentrums für Technomathematik für ein angenehmes Arbeitsumfeld bedanken. Insbesondere danke ich der Zerspangruppe und der SCiE-Gruppe für Fragen, Anregungen und Motivation. Ein besonderer Dank gilt dabei Dr. Kanglin Chen, Dr. Thanh Son Nguyen, Quy Muoi Pham und Majid Salmani, mit denen ich ein Büro teilen durfte.

Prof. Dr. Dirk Lorenz danke ich dafür, dass er mir mit seinen Erfahrungen fachlich und überfachlich zur Seite steht, mich inspiriert und motiviert, und ich danke Lena Liebsch für ihre Unterstützung, Geduld und Nachsicht.

Mein größter Dank gilt meinen Eltern Heidi und Carsten Kanning und meiner Schwester Katrin Lorenz. Ohne euch wäre ich nicht hier. Ihr habt mir die nötige Neugier und den Mut gegeben, alles zu hinterfragen!

Contents

List of Figures

1 Introduction

On its opening day in June 2000, the Millennium Bridge over the river Thames in London started to wobble alarmingly under the weight of thousands of people. Two days later the bridge was closed, cf. [SAM$^+$05]. Almost 60 years earlier a similar event gained popularity under the name of the *Tacoma Narrows Bridge disaster* or the *Galloping Gertie*. Only four month after its completion in July 1940 the suspension bridge in the state of Washington collapsed due to strong torsional vibrations caused by nothing but winds, as it is recalled by Gunns in [Gun81]. The increasing vibration of both bridges was a reaction to a periodic load transmission near a so-called eigenfrequency.

The scientific analysis of solids in motion, such as bridges or all kinds of buildings and mechanical systems, is known as *structural dynamics*. As the mentioned examples show, understanding the dynamical behavior of complex structures is still challenging, and reliable predictions can not be taken for granted. This thesis is ought to provide the foundation for a thorough approach to structural dynamics. Moreover, the aim is to show occurring obstacles from both the mathematical and the engineering point of view. Although the presented mathematics may also be valid for civil engineering scenarios as described above, the focus is on applications in mechanical engineering, especially production machines.

Deviations from an intended work process of e.g. cutting and turning machine tools are indicators of a machine's accuracy. Insufficient stiffness of a machine tool is only one cause of unsatisfactory precision. Due to

1

modern methods of calculation, the determination of the static stiffness of a machine during construction is possible with satisfying results. The analysis of dynamics on the other hand, especially for coupled mechanical systems, is still challenging because of too many or even unpredictable interactions. Imbalanced dynamic behavior of machines causes vibrations and thus not only bad surface quality of the workpiece, but also a higher abrasive wear and even broken die, as it is stated in [WB06].

The problem of controlling vibrations in production processes has occupied scientists for over a century, and modal analysis still is the most common tool for the estimation of the dynamical behavior of structures. An experimental modal analysis makes use of measured vibrational responses of a real mechanical structure to known load transmissions in order to explore the structure's dynamical behavior. One of the restrictions of the classical modal analysis is that load transmissions are either single wide-band impulses with an impact hammer, or mono-frequent excitations over an extend period of time, by e.g. an electromagnetic shaker. Responses to these kind of stimulations are said to be stationary, and modern production facilities often have various quickly changing load transmissions. Analyzing the dynamical behavior of an accelerating rotating drive may not be reasonable with a classical modal analysis or established mathematical tools, cf. [DMM+05].

Exploring the possibilities that arise from measuring vibrational responses in order to optimize work processes is the aim of this thesis. Under these circumstances, the *possibilities* are generally known as *inverse problems*. In order to eventually realize a potential optimization, the first step is a thorough theoretical analysis of simple mechanical systems. Therefore, adequate mathematical models have to be understood, and modeling has to be possible quickly and flexible. Obviously, a mathematical model has to be capable of providing simulations of the system's dynamical behavior. Applied methods and theoretical results are being published within many active research fields such as mechanical engineering, civil engineering, electrical engineering, signal processing, inverse problems, optimization and optimal control, just to mention a few.

A general introduction to the discussed applications and additional motivation can be found in "Werkzeugmaschinen" by Manfred Weck and Christian Brecher, cf. [WB06], a textbook about machine tools, especially about measurement-assisted analyses and dynamic stability. The textbooks "Strukturdynamik" volume 1 and 2 by Robert Gasch and Klaus Knothe contain a general introduction to structural dynamics from an engineering point of view, cf. [GK89a, GK89b]. Volume 2 also contains a brief but comprehensive introduction to the finite element method, which is a method for discretizing partial differential equations, and its application in structural dynamics that has its roots in structural dynamics, cf. [ZT94].

A wide array of literature has been published on the use of the finite element method in structural dynamics. Klaus Knothe and Heribert Wessels's "Finite Elemente" is meant as an introduction for engineers, cf. [KW99]. Olgierd C. Zienkiewicz and Robert L. Taylor pick up the topic from a structural dynamics point of view, but proceeds to cover the abstract mathematical side as well in [ZT94]. The connection of structural dynamics and the finite element method is also explicitly mentioned in [HK04, WJ87]. In [Gra04], Wodek K. Grawonski relates structural dynamics and control methods in order to create a common language for engineers from both fields. A comprehensive text on mathematical control theory is given by Eduardo D. Sontag in [Son98].

Inverse problems, related to structural dynamics, have also been tackled from various scientific positions with various definitions of what an inverse problem in this context is. In [Sim81a, Sim81b], Stepan S. Simonian discusses theory and application of parameter identification problems in structural dynamics, where he uses a frequency-domain approach and mathematical optimization. The problem of identifying parameters for a finite element model of a structure is known as model updating. Here, the works of Mottershead and Friswell give a good overview about the history and recent progress, cf. [MF93, FMA01]. The pure mathematical aspect of inverse problems in structural dynamics was supposed to be the construction of symmetric matrices with a band structure from given eigenvalues by Graham M. L. Gladwell in [Gla97].

Among the first attempts to account for ill-posedness in inverse problems in the field of structural dynamics is the article "Regularisation Methods for Finite Element Model Updating" by Hamid Ahmadian et al., cf. [AMF98]. More recent publications by Branislav Titurus and Michael I. Friswell, cf. [TF08] and Michael I. Friswell, cf. [Fri08] are also concentrated on the problem of model updating. As another aspect of inverse problems in the context of structural dynamics, the identification of load transmissions has to be mentioned. Dicken et al. and Ramlau et al. identify imbalances in rotordynamics in [DMM+05] and [RDM+06], respectively. Both approach the problem the frequency-domain.

The thesis is organized as follows. In the preliminaries, the basic mathematical theory for vibrational analyses and some additional results from functional analysis are recalled. The chapter starts with some conventions on frequently appearing notations. Chapter 3 is supposed to give an overview about the physical background of structural dynamics for simple mechanical systems. Therefore, basic concepts of vibrating solids and their simulation are being recalled. The concepts of chapter 3 are extended to more complex mechanical systems in chapter 4. In section 4.1, the application of the finite element method to discretize beam-like structures is introduced and exemplified. The computation of general vibrational responses is treated in the last section of chapter 4, while along the way some remarks on energy dissipation are given.

Chapter 5 contains the main contributions to the field of structural dynamics, presented within this thesis. In the beginning of the chapter, a brief introduction to the mathematical aspect of inverse and especially ill-posed problems is given. A mathematical approach to model updating is presented afterward in section 5.2. By means of model updating, a mathematical model of an existing structure may be improved through a fitting process with actual measurements. In section 5.3, a method for identifying arbitrary load transmissions from given measurements of vibrational responses is presented. The chapter closes with a brief introduction to input shaping in section 5.4, and the final chapter contains a summary and a list of open problems.

The research that lead to this thesis was partially performed in a collaboration with Linghan Li as a part of the graduate course "Scientific Computing in Engineering" at the University of Bremen. Especially the experimental results were gathered in the laboratories of the Bremen Institute for Mechanical Engineering at the University of Bremen (bime). The work of Linghan Li was supervised by Prof. Dr. Bernd Kuhfuß, head of bime, and will be submitted as a doctoral thesis presumably in 2013, cf. [Li13].

2 Preliminaries

The following preliminary words are meant to recall mathematical concepts that are necessary or may be helpful during the course of this thesis. The chapter contains relevant definitions and theorems, while the aim is not to provide an introduction to the theory. An introduction to the required concepts from physics and engineering sciences is given in the next chapter.

In this thesis, the word *system* appears frequently and may refer to 2 different things. The first being 2 or more connected solid bodies, such as metal beams, which will be called a *mechanical system*, the latter being 2 or more mathematical equations that can only be solved together, which will be called a *system of equations*. Whenever the meaning is clear from the context, the word system may also appear solely.

The natural numbers, starting from 1, are denoted by $\mathbb{N}$ and unless stated otherwise, n is an arbitrary element of $\mathbb{N}$. The real numbers are denoted by $\mathbb{R}$, the complex numbers by $\mathbb{C}$. Neither of these notations include infinities. A frequently appearing notation is $\mathbb{R}_0^+$, i.e. all positive real numbers including zero. An n-dimensional Euclidean space will be denoted by $\mathbb{R}^n$, and if not stated otherwise, all vector spaces are real. For two vector spaces X and Y, the space of continuous functions from X to Y is denoted by $C(X,Y)$, and for $n \in \mathbb{N}$ the space of functions from X to Y with continuous n-th derivative is denoted by $C^n(X,Y)$.

For the sake of completeness, some basic statements from functional analysis are cited, here. A thorough discussion can be found in most textbooks on the topic. If not stated otherwise, proofs can be found in, e.g. in [Wer05].

Definition 2.1

Let X and Y be real Hilbert spaces. The space $X^* := L(X, \mathbb{R})$ is called the *dual space of* X and $Y^* := L(Y, \mathbb{R})$ accordingly the dual space of Y. If $T \in L(X, Y)$, the operator $T^* : Y \to X$ is called *adjoint of* T and can be computed by showing that for all $x \in X$ and $y \in Y$

$$\langle Tx, y \rangle_X = \langle x, T^*y \rangle_Y.$$

Definition 2.1 can be directly transferred to the case, where X and Y are complex vector spaces by just substituting $\mathbb{R}$ by $\mathbb{C}$.

Proposition 2.1

Let M be a compact self-adjoint operator on a real or complex Hilbert space H. Let e_n be an orthonormal basis of H consisting of eigenvectors of M with corresponding eigenvalues λ_n. Then, the operator

$$\sqrt{M}x := \sum_{n=1}^{\infty} \sqrt{\lambda_n} \langle x, e_n \rangle e_n, \qquad (2.1)$$

called *square root of* M is well-defined, compact and self-adjoint. It holds $\sqrt{M}^2 = M$.

Proof. See [Ped00].

$\square$

A notation for appearing spaces of integrable function is given in the following definition.

Definition 2.2

Let $\Omega \subset \mathbb{R}^n$ be a measurable subset of a real vector space. For $0 < p < \infty$ the space of all measurable functions $f : \Omega \to \mathbb{R}^n$, for which the L^p-*norm*

$$\|f\|_{L^p} := \left(\int_\Omega |f(x)|^p dx \right)^{\frac{1}{p}}$$

is finite, is denoted by $L^p(\Omega, \mathbb{R}^n)$. Functions that are identical almost everywhere are not told apart in an L^p-space.

An equal definition can also be made for complex vector spaces and $p = \infty$ without any restrictions, cf. [Wer05]. Unless stated otherwise, all appearing integrals are meant component wise, e.g. for $f \in L^p(\mathbb{R}, \mathbb{R}^2), f(x) = [f_1(x), f_2(x)]^T$ is understood as

$$\int_{\mathbb{R}} f(x)dx = \begin{bmatrix} \int_{\mathbb{R}} f_1(x)dx \\ \int_{\mathbb{R}} f_2(x)dx \end{bmatrix}.$$

Above and in the following, a superscript T denotes a transpose. The following famous *Hölder inequality* is essential when dealing with L^p-spaces.

Theorem 2.2

Let $1 \leq p < \infty$, and let $f \in L^p, g \in L^q$, such that

$$\frac{1}{p} + \frac{1}{q} = 1.$$

Then, it holds

$$\|fg\|_{L^1} \leq \|f\|_{L^p} \|g\|_{L^q}.$$

Proof. See [Wer05]. □

The L^p-spaces must not be mistaken for the space of bounded linear operators between the spaces X and Y, denoted by $L(X, Y)$. Whenever the domain and range of a mapping are clear from the context, their indication may be omitted. In general, the domain of a mapping f is denoted by $\mathrm{dom}(f)$ and its range by $\mathrm{ran}(f)$, i.e. $f : \mathrm{dom}(f) \to \mathrm{ran}(f)$.

The real L^2-space equipped with the inner product

$$\langle f, g \rangle_{L^2} := \int_{\mathbb{R}} f(x)g(x)dx$$

is a Hilbert space. A useful and well-known integral transform on that space can be defined as follows.

Definition 2.3

Let $f \in L^2(\mathbb{R}^n)$. Define

$$(\mathscr{F}f)(\omega) = \frac{1}{(2\pi)^{n/2}} \int_{\mathbb{R}^n} f(x)e^{-ix\omega}dx, \tag{2.2}$$

for all $\omega \in \mathbb{R}^n$. Then, $\mathscr{F}f$ is called the *Fourier transformed of f* and the mapping $\mathscr{F}$ is called *Fourier transform*.

It is obvious that $\mathscr{F}f$ is well-defined and measurable, and that $\mathscr{F}$ is linear. Usually, the Fourier transform is first defined for L^1-functions only, and showing the validity for L^2 functions requires some additional thoughts, which can be found e.g. in [Wer05].

Remark 2.3

The Fourier-transform is a continuous bijection on L^2.

A special feature of the Fourier transform is its impact on derivatives.

Proposition 2.4

For $\Omega \subset \mathbb{R}^n$ let $f \in L^2(\Omega, \mathbb{R}^n)$ and let α be a multiindex, i.e. a set of non-negative integers $\alpha_1, \ldots, \alpha_n$ such that $|\alpha| = \sum_i \alpha_i$. Then with

$$D^\alpha := \frac{\partial^{|\alpha|}}{\partial^{\alpha_1} x_1 \cdots \partial^{\alpha_n} x_n}$$

it holds

$$\mathscr{F}(D^{(\alpha)}f)(\omega) = i^{|\alpha|}\omega^\alpha \mathscr{F}f(\omega).$$

Proof. See [Wer05]. $\qquad\qquad\square$

The reader, not familiar with the concept of derivatives in function spaces is referred to [Wer05] for further explanations.

For functions $f, g \in L^p(\mathbb{R}, \mathbb{R}^n)$ the *convolution of f and g* is defined as

$$(f * g)(t) = \int_{\mathbb{R}} f(t-s)g(s)ds,$$

and it is well-defined and measurable, cf. [Wer05].

Lemma 2.5

Let $f, g \in L^2(\mathbb{R}, \mathbb{R}^n)$, then

$$\mathscr{F}(f * g)(\omega) = \mathscr{F}f(\omega)\mathscr{F}g(\omega).$$

Proof. See [Wer05]. □

Another important result about convolutions is known under the name of *Young's inequality.*

Theorem 2.6

Let $1 \leq p, q < \infty$ sich that

$$\frac{1}{r} := \frac{1}{p} + \frac{1}{q} - 1$$

is non-negative. If $f \in L^p(\mathbb{R}, \mathbb{R}^n)$ and $g \in L^q(\mathbb{R}, \mathbb{R}^n)$, then $(f * g) \in L^r(\mathbb{R}, \mathbb{R}^n)$ and it holds

$$\|f * g\|_{L^r} \leq \|f\|_{L^p}\|g\|_{L^q}.$$

Proof. See [Wer05]. □

Finally, two notations of abbreviating character will be used within this thesis. Since functions frequently depend on a time variable t and a space variable x, derivatives towards t are denoted by dots, and derivatives towards space are denoted by primes, e.g. for $f \in C^2(\mathbb{R}, \mathbb{R})$,

$$\dot{f}'(t, x) := \frac{\partial^2}{\partial t \partial x} f(t, x),$$

$$\ddot{f}(t, x) := \frac{\partial^2}{\partial t^2} f(t, x),$$

$$f''(t, x) := \frac{\partial^2}{\partial x^2} f(t, x).$$

3 One Degree of Freedom

The number of *degrees of freedom* (DOF) of a mechanical system counts its possibilities to shift and turn in space. For a rigid body, it consists of the number of possible movements in 3 different spatial dimensions plus its momenta around 3 rotational axes and thus is greater or equal to zero and less or equal to 6. Every real solid is assumed to have infinitely many DOF, if some sort of flexibility is accounted for. Within this chapter, a system with only 1 DOF will exemplify the issues of vibrational analysis and its benefits. This might appear restrictive, but 1 DOF is enough to explain the main conclusions about oscillating linear systems, and even the dynamical behavior of systems with n DOF can always be traced back to n systems with 1 DOF each, cf. [GK89a] or [WJ87].

In the first section, a general mathematical equation and its solution will be introduced that suffices to describe the dynamical behavior of mechanical systems. This equation will be augmented with an additional term to include loss of energy over time in the second section. In the final section, numerical and experimental results from a realization of a system with 1 DOF are presented. Most of the content within this chapter may be helpful as an additional introduction for mathematicians, while it may serve as a recall for mechanical engineers. The presentation mainly relies on the textbooks [GK89a, WJ87, WB06] and the article [TM01].

3.1 Dynamical behavior

In order to model the dynamical behavior of an elastic structure, its mass has to be known and its stiffness has to be derived from the structure's ability to hold static loads. The stiffness s of a spring is given by the constant, relating an applied force F to an extension or compression Δl in longitudinal direction,

$$s = \frac{F}{\Delta l}. \tag{3.1}$$

Conventionally, force F is measured in Newton, denoted by $[N]$, and variations Δl in length are measured e.g. in meter $[m]$, as it can be found in textbooks on physics and engineering, see [Ger01, GHS03]. The stiffness of a spring, the so called spring constant c can be determined experimentally with a spring scale. At the free end of a cantilever beam, its stiffness b is given by

$$b = \frac{3EI}{l^3}, \tag{3.2}$$

where l is the length of the cantilever, E is a material property, called elastic modulus and I is the so called moment of inertia of the cross section, see [WJ87]. More insight on this thought is given in definition 4.1 and in chapter 4. The cross-sectional moment of inertia is a geometric property and also known as the second moment of area of the cross section, as it is stated in [SGH02].

In [GK89a], Gasch and Knothe explain the origin of the equation, describing variations in 1 DOF by considering a pendulum with positive mass m and length l, which is displaced over time t from its position of rest by the angle function $\varphi \in C^2(\mathbb{R}^+, \mathbb{R})$. The acceleration of the mass is given by the second derivative of the displacement in time $l\ddot{\varphi}(t)$, since the arc length of the pendulum is $l\varphi(t)$. The restoring force is given by $mg \sin \varphi(t)$, with standard gravity g. It works in opposite spatial direction of the displacement. Therefore, these forces have opposing signs. The mathematical equation can be derived from Newton's second law, which says that force equals mass times acceleration

$$ml\ddot{\varphi}(t) = -mg \cdot \sin \varphi(t)$$
$$\Longleftrightarrow \quad ml\ddot{\varphi}(t) + mg \cdot \sin \varphi(t) = 0,$$

with an initial condition $\varphi(t_0) = \varphi_0$. When dealing with vibrations of rather stiff mechanical systems, where the angle $\varphi(t)$ is close to zero, Gasch and Knothe suggest to approximate $\sin \varphi(t)$ by the angle itself, giving the second order ordinary differential equation (ODE)

$$ml\ddot{\varphi}(t) + mg \cdot \varphi(t) = 0. \tag{3.3}$$

Two additional attributes of (3.3) may be noted. All terms depend linearly on the solution φ, and thus, the ODE is homogeneous and linear.

Weaver and Johnston deduce equation (3.3) from an application of D'Alembert's principle, cf. [WJ87]. For a given elastic structure, the general variation over time of 1 DOF is denoted by the scalar valued function $u \in C^2(\mathbb{R}^+, \mathbb{R})$ and thus, its acceleration is $\ddot{u}(t)$. With positive scalars mass m and stiffness s, determined from the structure properties, a restoring force $su(t)$ and an inertial force $m\ddot{u}(t)$ occur. Both forces act against the direction of displacement, giving

$$-su(t) - m\ddot{u}(t) = 0$$
$$\Longleftrightarrow \quad m\ddot{u}(t) + su(t) = 0. \tag{3.4}$$

This equation and equation (3.3) are called homogeneous equation of motion. With a system of equations of this type it is possible to describe variations and interactions of several DOF simultaneously. One way of modelling mechanical systems with multiple DOF will be explained in section 4.1.

A solution of this ODE may be obtained from the ansatz

$$u(t) = u_0 e^{\omega_0 t}. \tag{3.5}$$

In order to obtain $u_0 \in \mathbb{R}$, initial conditions for the ODE must be formulated and $\omega_0 \in \mathbb{C}$ is the solution to

$$\omega_0^2 m + s = 0 \tag{3.6}$$

$$\implies \quad \omega_0 = \pm\sqrt{-\frac{s}{m}}. \tag{3.7}$$

The solution ω_0 describes one of the main features of structural dynamics. The following definition provides a more general point of view for equation (3.6) and is useful for the characterization of mechanical systems.

Definition 3.1

Let M, C and K be complex matrices of rank n. Computing the eigenvalues λ_i and the corresponding right and left eigenvectors v_i and w_i, respectively, for $i = 1, \ldots, 2n$ from

$$(\lambda^2 M + \lambda C + K)x = 0 \quad \text{and} \quad y(\lambda^2 M + \lambda C + K) = 0,$$

is called a *quadratic eigenvalue problem* (QEP).

Thus, equation (3.6) is a special QEP, where $M = m$ and $K = s$ are real and of rank 1, while C is equal to zero. Recently, QEPs received special attention. A comprehensive review on how they are solved and the different fields, they occur in is given by Françoise Tisseur and Karl Meerbergen in [TM01]. Different from the standard and generalized eigenvalue problems,

$$Ax = \omega_s x \quad \text{and} \quad Ax = \omega_g Bx,$$

no canonical form, equivalent to the Schur decomposition exists for the QEP. The QEP (3.6) and its solution ω_0 have a special meaning to mechanical systems, which is worth noting in the following definition.

Definition 3.2

Let m and s be positive real numbers. The imaginary part of the solution of the QEP (3.6), given by

$$\omega_0(m, s) := \sqrt{\frac{s}{m}}, , \tag{3.8}$$

is called *natural frequency of m and s.*

In general, the term frequency is used to describe periodic dynamcial behavior, and an array of literature on so called time-frequency analysis is available, e.g. [Gro01]. One of the goals in time-frequency analysis is an optimal representation of the combined time- and frequency-behavior of a signal. Plotting a signals amplitude against time gives the best time resolution, but no direct information about frequencies[1]. The Fourier transform of a signal on the other hand qualitatively represents "energy per frequency", which admits the following definition.

Definition 3.3
Let $f \in L^2$. Then, $\mathscr{F}f$ is also called *frequency spectrum of f*, and the modulus $|\mathscr{F}f|$ is called *amplitude spectrum of f*.

For a measured vibrational response of an ideal system with 1 DOF, the maximum of the signal's amplitude spectrum coincides with the systems natural frequency.

In the idealized case, where all applied energy is preserved within the system, any finite load transmission will result in an everlasting oscillation at $\omega_0(m, s)$Hz after the excitation has ended. The duration T of 1 period of oscillation is given by

$$T = \frac{2\pi}{\omega_0(m, s)}. \tag{3.9}$$

In reality, a perpetual motion machine cannot exist, which means that energy will dissipate during this oscillation. This is also stated as the first law of thermodynamics and can be found e.g. in Baehr and Kabelac's textbook

---

[1] Counting oscillations per second is sometimes possible

[BK09]. A closer look on the dissipation of energy is given in the next section and in section 4.43.

Definition 3.4

For $f_0, \omega \geq 0$ define

$$f : \mathbb{R}^+ \to \mathbb{C}, \quad f(t) := f_0 e^{i\omega t}. \tag{3.10}$$

Then, f is called a *harmonic load*, and the solution $u \in C^2(\mathbb{R}^+, \mathbb{C})$ of the inhomogeneous ODE

$$m\ddot{u}(t) + su(t) = f(t) \tag{3.11}$$

is called *stationary*.

For positive f_0 and ω, let $f(t) := f_0 e^{i\omega t}$ be a harmonic load, acting on the system described by equation (3.4). The system is now literally forced to follow this excitation [GK89a], and the ansatz $u(t) = u_0 e^{i\omega t}$, yields

$$m\ddot{u}_0 e^{i\omega t} + su_0 e^{i\omega t} = f_0 e^{i\omega t}$$

$$\Longleftrightarrow \quad \left(-\omega^2 + \frac{s}{m}\right) u_0 e^{i\omega t} = \frac{f_0}{m} e^{i\omega t}.$$

The solution u, describing the systems reaction to the load transmission does not depend on time anymore, and it holds

$$u_0 = \frac{f_0}{m\left(\frac{s}{m} - \omega^2\right)}. \tag{3.12}$$

If ω approaches the systems natural frequency, u_0 will approach infinity.

In mechanical engineering, excitations of a structure near the natural frequency ought to be circumvented to ensure its stability. In order to meet customers's expectations in all kinds of vehicles, the amount of occurring vibrations has to be controlled, see [DMM+05, RDM+06]. The accuracy of chip removal tools with a rotating drive may suffer severely from a turning speed close to a natural frequency. In electrical engineering on the other hand, resonance might be wanted explicitly, e.g. for tuning a radio [TM01]. It is now obvious that the natural frequency of a system is one of its main characteristics to understand its dynamical behavior.

3.2 Dissipation of energy

The dynamic behavior of mechanical systems, discussed until now, is the afore mentioned idealized case, in which all energy, induced to the system is preserved. With a model like this, responses to load transmissions with a finite duration will always become stationary after the excitation has ended. In a real structure, transmitted energy is converted into heat due to friction, cf. [WJ87]. For an analysis of instationary dynamical behavior of a mechanical system it is therefore necessary to include the loss of energy within the mathematical model. In the following, 2 different ways of determining the dissipation of energy will be introduced. In both cases a so called viscous damping is computed, where the energy is assumed to decrease proportional to the systems velocity, cf. [WB06, WJ87, GK89a]. The general inhomogeneous, damped equation of motion is given by

$$m\ddot{u}(t) + c\dot{u}(t) + su(t) = f(t), \tag{3.13}$$

where m, c and s are positive real numbers and u and f is a complex valued function meeting some smoothness constraints. In [Zha90, Zha94], Zhang analyzed general nonlinear dissipation of energy, and an explicit attempt to model a so called squeeze film damper which exhibits nonlinearity is presented by Dicken et al. in [DMM+05].

To obtain a solution for equation (3.13) the QEP

$$\lambda^2 m + \lambda c + s = 0 \tag{3.14}$$

has to be solved with positive real numbers m, c and s.

Definition 3.5
Let m, c and s be positive real numbers, and let $\omega_0(m, s)$ denote the natural frequency of m and s. The *damping ratio according to Lehr* is given by

$$D(m, c, s) := \frac{c}{2\omega_0(m, s)m}, \tag{3.15}$$

and the imaginary part of the solution of the QEP (3.14), given by

$$\lambda(m, c, s) = \omega_0(m, s)\sqrt{1 - D(m, c, s)^2},$$

is called *eigenfrequency of m, c and s*.

In [WB06], it is stated by Weck and Brecher that although the amount of energy which is lost over time separates the eigenfrequency from the natural frequency, the variation is too small, to be measured for real mechanical systems. The damping ratio according to Lehr is mentioned, e.g. in [GK89a].

In the case, where some measurement of the dynamical behavior of a mechanical system can be displayed as a decaying sine function, a very basic way to estimate the dissipation of energy may be applied as it is shown by Weck and Brecher in [WB06] or Gasch and Knothe in [GK89a].

Lemma 3.1

Given the values $u_0 = u(t_0)$ and $u_n = u(t_0 + 2n\pi), n \in \mathbb{N}$ of a decaying sine function $u \in C^\infty(\mathbb{R}^+, \mathbb{R})$,

$$u(t) := e^{-ct}\sin(t).$$

Then, the viscous damping coefficient c is given by

$$c = \frac{1}{2n\pi}\ln\left(\frac{u_0}{u_n}\right).$$

Proof. It holds

$$\frac{u_0}{u_n} = e^{c2n\pi}\frac{\sin(t_0)}{\sin(t_0 + 2n\pi)}$$

and thus

$$\ln\left(\frac{u_0}{u_n}\right) = 2\pi c n.$$

$\square$

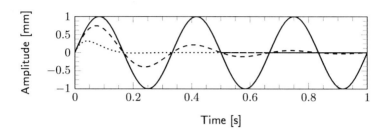

Figure 3.1: Three signals with different damping ratios in the time domain.

A change in frequency of the sine function by the factor ω, i.e. $u(t) :=$ $e^{-c\omega t}\sin(\omega t)$ results in

$$\ln\left(\frac{u_0}{u_n}\right) = 2\pi c n\omega.$$

With this remark, the dissipation of energy within a vibrational response of a mechanical system can be estimated by the ratio of 2 local maxima of its amplitude. With a printed signal at hand, local maxima and the according amplitude can be compared easily. The impact of the damping coefficient c is visualized in figure 3.1 with numerically computed values of

$$u_i(t) = e^{-c_i 3t}\sin(3t), \tag{3.16}$$

discretized in time at 3kHz. The 3 different damping coefficients are set to $c_1 = 0$, $c_2 = 0.2$ and $c_3 = 0.5$. Figure 3.1 qualitatively shows u_1 as a smooth line, u_2 as a dashed line and u_3 as a dotted line.

Another method to determine damping ratios involves and analysis of the oscillation's amplitude spectrum. It takes advantage of the fact that as the damping ratio decreases, the peak in the amplitude spectrum near its eigenfrequency gets higher, steeper and slightly shifted towards infinity.

Definition 3.6

Let $[m, c, s] \in \mathbb{R}^3$ describe a damped mass-spring system, excited by $f \in L^2(\mathbb{R})$, as in (3.13). Then, the function $G : (\mathbb{R}^3, L^2) \to L^2(\mathbb{R}^+, \mathbb{C})$, given by

$$G([m, c, s], f) := \frac{\mathscr{F}u}{\mathscr{F}f},$$

where $u \in \mathbb{C}^2$ is the solution of (3.13), is called *receptance of m, c, s and f*.

It holds

$$G(u, f)(\omega) = \frac{(\mathscr{F}u)(\omega)}{(\mathscr{F}f)(\omega)} = \left(-\omega^2 m + i\omega c + s\right)^{-1}.$$

The receptance $G(\cdot, \cdot)(\omega)$ represents the displacement of 1 DOF per unit force at a frequency ω.

Lemma 3.2

Let $u \in L^2(\mathbb{R})$ be a solution of (3.13) and let $G([m, c, s], f)$ be the receptance of m, c, s and f. If $\omega_0 \in \mathbb{C}$ is the maximum of $G([m, c, s], f)$, choose $\omega_L < \omega_0$ and $\omega_R > \omega_0$, such that

$$G([m, c, s], f)(\omega_L) = G([m, c, s], f)(\omega_R) = \frac{1}{\sqrt{2}} G([m, c, s], f)(\omega_0).$$

Then, the viscous damping coefficient in (3.13) is given by

$$c = m(\omega_L - \omega_R).$$

Remark 3.3

The above lemma is stated without proof. It is based on the so called half-power bandwidth method, as it is described in [WB06] or [OR10]. The damping ratio according to Lehr can be obtained without explicit knowledge of the value m. Given ω_0, ω_L and ω_R as defined above, it holds

$$D = \frac{(\omega_R - \omega_L)}{2\omega_0}. \tag{3.17}$$

Figure 3.2: The experimental setup for a system with 1 DOF and a modalhammer

3.3 Experimental results

The final section of this chapter contains results from a vibrational analysis of a realization of a system with 1 DOF, also referred to as an oscillating point mass. The purpose of the following is to verify the theory and conclusions of the above sections. The simple experimental setup, shown in figure 3.2, suffices as an example. The figure shows an aluminum cube, attached to a steel spring. Measurements of the vibrational responses were taken with an accelerometer which is also visible in figure 3.2, and the image on the right shows a modal hammer. This tool is equipped with a piezoelectric sensor that allows to estimate the transmitted force. The experimental setup was realized at the laboratory of the bime, and the experiments were performed in a collaboration with Linghan Li.

In order to cope for the assumption of an oscillating point mass, only displacements of the cube, longitudinal to the spring were taken into account. In order to simplify the system further, the mass of the spring ($m_s = 0.014$kg) and accelerometer ($m_a = 0.004$kg) were added to the mass of the cube ($m_c = 0.25$kg). Moreover, the stiffness of the aluminum cube is large enough, compared to the stiffness of the spring, to justify being con-

sidered rigid. Thus, the model can be seen as a system with 1 DOF, and
its dynamical behavior is governed by the equation (3.13) with appropriate
m, c and s. The system has been excited with single or double impulses,
manually transmitted with a modal hammer.

In comparison with an actual measurement, an according simulation, computed by solving (3.13) for a known excitation g, reveals a major flaw of the
model, see figure 3.3. The example clearly shows that a crucial amount of
the excitation is contained within the measured vibrational response, while
this is not the case for the simulation. It seems that the model is only
representing the dynamical behavior of the system before and after the impulses. For now, this obstacle is cleared by subtracting a scaled amount of

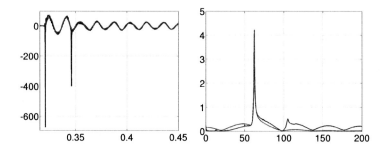

Figure 3.3: Measurement (blue) and simulation (red) of a vibrational response (left) and according amplitude spectra (right)

the excitation after computing the vibrational response from (3.13), see e.g.
figure 3.5. Further thoughts on this observation will be presented in the last
chapter.

From close-ups on the measured impulses it can be suggested that the
actually transmitted impulses are more adequately modeled with a sharply
located Gaussian than with a Dirac-impulse. This observation motivates
defining load transmission of *impulse-type*. A load transmission of impulse-

type is understood as a function $f \in L^2(\mathbb{R}, \mathbb{R}^n)$,

$$f(t) = f_0 e^{-\left(\frac{t-t_0}{\sigma}\right)^2}, \tag{3.18}$$

where f_0 represents the force and t_0 the time of the impact, and $\sigma << f_0$. Figure 3.4 shows a measured impulse (blue) and a fitted load transmission of impulse-type (red).

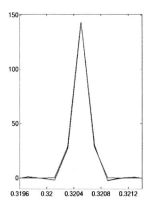

Figure 3.4: Measured impulse (blue) and load transmission of impulse-type (red)

3.3.1 Measurements

The parameter m is known immediately. The sum of m_s, m_a and m_c gives a total mass of $m = 0.268$kg. The static stiffness of the spring has been measured with a spring scale. The approximate value, resulting from this measurement was $s_s = 43\frac{\text{kN}}{\text{m}}$. By applying the theoretical results from the previous sections, the parameters c and s may also be determined from measured vibrational responses. As the name indicates, the attached accelerometer measures acceleration in a certain spatial direction. This circumstance leads to a small inconvenience. All formulas above refer to the displacement in a DOF and thus, simulations have to be derived twice to be comparable with the measurements. Additionally, as mentioned above, the forced displacement due to the impact of the modal hammer is not part of the vibrational response. The sampling rate was set to 5120Hz.

For given mass m, as well as measurements of the exciting force f and the resulting acceleration A_m of the structure in 1 spatial direction, the structures stiffness s can be estimated by

$$ s = \frac{2\pi m \omega^2}{\sqrt{1 - D^2}}, \tag{3.19} $$

where the eigenfrequency $\omega = 61.88$Hz is taken as the maximum of the receptance, known as the *peak amplitude method*, cf. [BG63]. The quantity D is an estimation of the damping, obtained by the half-power bandwidth method, see lemma 3.2 and remark 3.3. In order to estimate the damping ratio D, lemma 3.1 may also be applied. Given the values ω, u_0, u_n and n, from the printed dynamical response of the system, definition 3.15 of the damping ratio according to Lehr, yields

$$ c = \frac{1}{2\pi n \omega} \ln\left(\frac{u_0}{u_n}\right), $$

$$ D = \frac{1}{mn} \ln\left(\frac{u_0}{u_n}\right). $$

Given a measurement of the vibrational response to an impulses f, application of (3.19) and remark 3.3 yields

$$ s = 40,537.728\,\frac{\text{N}}{\text{m}}, \quad \text{and} \quad c = 1.075. $$

Thus, the dynamical behavior of the system is governed by

$$ 0.268\ddot{u}(t) + 1.075\dot{u}(t) + 40,537.728u(t) = f_1(t). \tag{3.20} $$

An implementation of an explicit one-step method[1] was used to solve the ODE (3.20). For a more detailed discussion about one-step methods, see section 4.2.1, or [SB92, HrW87, Ise98]. The whole process of estimating the model parameters and solving the ODE was automated in a Matlab routine and performed on several data sets. Therefore it was necessary to approximate the measured impulses by Gaussians and to solve the ODE in at least 2 steps, depending on the number of impulses. The instants t_i of each impulse were extracted from the data, and the ODE was solved on the subintervals $[0, t_1], [t_1, t_2], \ldots$ until the end of the measured interval, where the final values of each solution was passed on as the initial value of the next ODE.

Figure 3.5 shows a measured vibrational response of the system to an impulse excitation, denoted as data set *1a*, and its computed approximation in the time domain. The according acceleration-amplitude spectrum (AAS) was computed with an implementation of the fast Fourier transform (FFT) and is displayed in figure 3.6. In both figures, the top graphic displays the measurement and the bottom one shows the simulation. The slight deviance in the simulation can be explained with the presence of noise in the measurement, and the fact that the experimental setup is only an idealized assumption of an oscillating point mass. As mentioned before, a scaled amount of the excitation was subtracted after computing the vibrational response.

The same procedure was done for another single impulse excitation response, denoted as data set *1b*, 2 double impulses and a triple impulse excitation, denoted as data sets *2a,2b* and *3*, respectively. The results of the according model parameter estimation are presented in table 3.1. It contains the estimated viscous damping factor, dynamic stiffness in N/m and eigenfrequency in Hz. The energy, transmitted to the structure is given in the column, entitled "energy". Here, the proportional relation between the

---

[1] The applied method was the Matlab routine ode23tb.

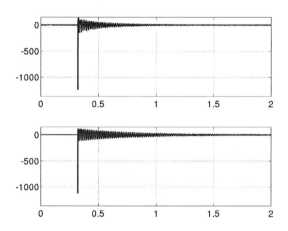

Figure 3.5: Measurement (top) and simulation (bottom) of a vibrational response in the time domain of data set 1a

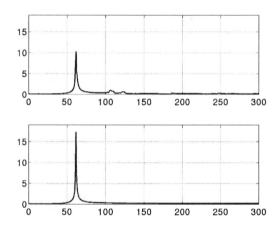

Figure 3.6: Acceleration-amplitude spectra of a measured vibrational response (top) and a simulation (bottom) of data set 1a

stiffness and the eigenfrequency can be seen, and it seems that these characteristics are proportional to the transmitted energy.

Data	Energy	Damping	Stiffness	Eigenfrequency
1a	2.140	1.075	40,537.728	61.883
1b	1.827	0.949	41,770.077	62.820
2a	2.009	1.119	40,949.764	62.195
2b	2.039	0.790	40,938.198	62.195
3	1.779	0.867	41,352.880	62.508

Table 3.1: Estimated model parameters

Additionally to the display of data set 1a, its AAS and the according simulations in figures 3.5 and 3.6, respectively, an equivalent presentation of data set 1b is displayed in figures ?? and ??, respectively. The data sets 2a, 2b and 3, their respective AAS and according simulations are displayed in figures 3.7 and 3.8, figures 3.9 and 3.10 and figures 3.11 and 3.12.

3.3.2 Estimating model parameters and output-only analysis

It is now verified, that equation (3.13) suffices as a mathematical model for simple mechanical systems, and that the methods for estimating the parameters for such a model, presented above are applicable and valid under stated conditions. Nevertheless, 2 drawbacks of those methods are the motivation for a broad field of research. First, the receptance is the input-output ratio in the frequency domain and demands knowledge of the input, which might not always be accessible since measuring is time-consuming, and sensors are expensive. Therefore, the possibility to estimate model parameters by so called *output-only* methods, i.e. methods that do not rely on measurements of the exciting force, are explored. Second, the operating conditions, the mechanical system is exposed to in reality, may differ significantly from those applied during the modal test. Model parameters may vary rapidly in a

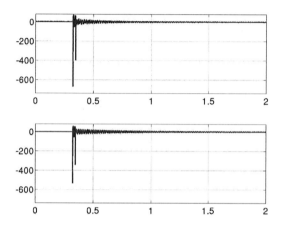

Figure 3.7: Measurement (top) and simulation (bottom) of a vibrational response in the time domain of data set 2a

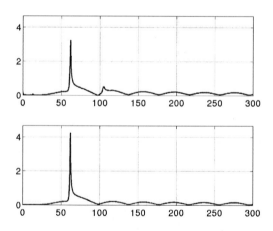

Figure 3.8: Acceleration-amplitude spectra of a measured vibrational response (top) and a simulation (bottom) of data set 2a

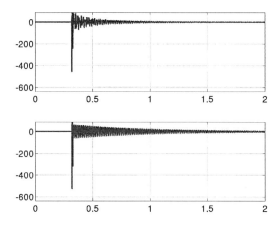

Figure 3.9: Measurement (top) and simulation (bottom) of a vibrational response in the time domain of data set 2b

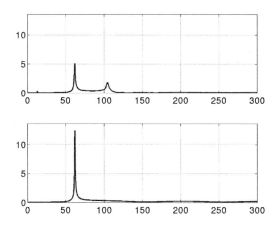

Figure 3.10: Acceleration-amplitude spectra of a measured vibrational response (top) and a simulation (bottom) of data set 2b

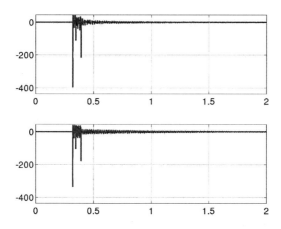

Figure 3.11: Measurement (top) and simulation (bottom) of a vibrational response in the time domain of data set 3

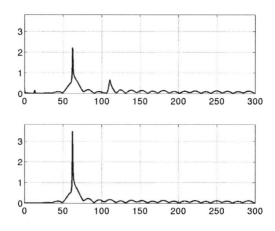

Figure 3.12: Acceleration-amplitude spectra of a measured vibrational response (top) and a simulation (bottom) of data set 3

short time due to variations in the operational parameters of machines and changes in boundary conditions, as it is stated by Spiridonakos and Fassois in [SF09]. Hermans and Van der Auweraer address these issues in an article on the analysis of structures under operational conditions, cf. [HdA99]. In a so-called *operational modal analysis*, measurements are made under real operational conditions, e.g. during regular production processes, cf. [MR04].

The canonical tool to determine a signal's frequency is the Fourier transform or a short-time Fourier transform, also known as the Gabor transform. Even if no knowledge of the excitation is needed, these methods are strictly limited when it comes to resolutions of a combined time-frequency representation, due to Heisenberg's uncertainty principle, cf. e.g. [MSSS10]. Thus, rapid variations of the model parameters can not sufficiently be identified with methods based on the Fourier transform. Motivated by the need to identify changing model parameters, various methods have recently been proposed. A brief survey and a comparison of 3 different output-only approaches for the identification of rapidly varying model parameters has been presented by Li et al. in [LKSK11].

The comparison was made between an unregularized least-squares based approach (LSE), proposed by Yang and Lin in [YL05], an estimation of signal parameters via rotational invariance techniques proposed in [PRK85] by Paulraj et al., also known as ESPRIT, and a maximum correlation approach, based on the so called orthogonal matching pursuit (OMP), which was originally proposed in the signal processing context by Davis et al., see [DMZ94]. All 3 methods are able to estimate a signal's resonance frequency and damping ratio. In [LKSK11], results of examinations of the estimation quality of these methods were presented. The experiments were performed at the bime in cooperation with the author of this thesis. The performance of all 3 algorithms was tested under varying noise levels and signal lengths for 2 different numerically simulated vibrational response signals with a fixed resonance frequency and damping ratio. The performance of the LSE approach was also examined for a numerically simulated vibrational response signal with a linearly changing resonance frequency and damping ratio.

The experiments showed the identification accuracy of the LSE algorithm to be higher than the ESPRIT algorithm, and that both methods exhibited higher identification accuracy than the OMP approach. As expected, the error in identifying the damping ratio is larger than in identifying the frequency for all tested methods. Additionally, estimating the damping ratio is more sensitive to the signal to noise ratio. A higher number of oscillation periods and reduced noise level in the signal data improved the identification accuracy. The final test showed that the LSE approach is capable of detecting changes in frequency and damping ratio. As opposed to the tested LSE method, ESPRIT and OMP are capable of estimating multiple resonance frequencies and damping ratios within a signal.

4 The Forward Problem

The deviation from the intended work process of e.g. turning and milling machine tools is an indicator of the machines accuracy. If a desired tool-path is optimally realized, the result may still not be satisfactory due to deflections and oscillations of the tool-tip. In order to analyze such a process theoretically, two things are needed. A model that contains all necessary information about the geometry and materials of the machine has to be generated, and a tool to predict its dynamical behavior, such as vibrations and deformations as reactions to load transmissions, has to be found. The first section in this chapter is concerned with expanding the concepts of the previous chapter. The aim is to describe geometries and materials of a mechanical system with multiple, coupled DOF by equations. Figure 4.1 shows an ultra-precise milling machine. The tool removes chips with optical precision, i.e. an accuracy of less than one nanometer. It will be shown, how these kind of structures can be modeled by a system of linear ODE,

$$M\ddot{u}(t) + D\dot{u}(t) + Su(t) = f(t), \tag{4.1}$$

where M, D and S are real matrices of rank $s \in \mathbb{N}$ describing the mass, damping and stiffness of the system's components. The function $f \in L^2(\mathbb{R}_0^+, \mathbb{R}^s)$ may contain external forces applied to the structure, and $u \in C^2(\mathbb{R}_0^+, \mathbb{R}^s)$ describes the displacement of the structure in s DOF at time t. As in the previous chapter, the incorporation of energy dissipation is treated separately. In section 4.2, it is first recalled, under which assumptions (4.1) has a unique solution. Tools to solve the ODE are presented afterward in the same section.

Figure 4.1: DMG ultra-precise milling machine

The s-dimensional system matrices, considered here are symmetric and of full rank and may thus be diagonalized by Gram-Schmidt, cf. [Bos08]. This process is called modal decoupling in the engineering science, cf. [GK89a, TM01] and it verifies the statement from chapter 3 that mechanical systems with s DOF can be seen as s systems with 1 DOF, each.

4.1 The finite element method for beam-like structures

In general, all mechanical systems and structures, such as production machines or simple construction elements, as shown in figure 4.1, may be seen as mass-spring systems, since vibrational analyses mainly depend on their mass and spring properties. The term damped mass-spring system is also used, if the model incorporates energy dissipation. Real structures always exhibit crucial damping ratios, but modeling a structure with the finite element method yields no damping matrix. According to Gasch [GK89a] or Tisseur

[TM01], damping properties of real structures are rarely known and sometimes difficult to evaluate precisely. One way of approaching the simulation of a damped structure will be presented in section 4.43. Within this section, the generated model will not include any damping. The starting point within this section is the formulation of a mathematical equation, describing the dynamics of elastic deformations of solid, homogeneous beams. Therefore, some basic results from physics and Lagrangian mechanics are being recalled. The main source for this section was the textbook "Strukturdynamik Band 2" by Robert Gasch and Klaus Knothe, see [GK89b]. Some of the concepts are cited from other sources, such as [GHS03, SGH02, KW99, HK04, Ger01], where additional insight to the topic of this section and its surroundings are provided.

4.1.1 Homogeneous and shear rigid beams

For matters of convenience, an agreement on the vocabulary, regarding beams may be made. Figure 4.2 shows a sketch of a beam element, all 6 DOF at a free end of a beam and the notation of the axes and displacements that is used in the following. The double arrowhead is a notation used in engineering literature, and it indicates a moment around the according axis. Moments around the x-axis are called *torque* and result in twists, denoted by β_x. Moments around an imaginary axis through the cross-section parallel to y, denoted by β_y, force the beam to arch in z-direction and vice versa. Together with shifts parallel to y and z, denoted by w_y and w_z, respectively, these lateral displacements are called *bending*. Pure shifts of the cross-section towards x are denoted by w_x. If they result in a change of length, they are called *stretching*.

Throughout this thesis, two assumptions on the geometry and material of solid beams are made. All solid beams are assumed to be *homogeneous*, i.e. their y-z-cross-section area and material properties are constant. Moreover, all beams are assumed to be *shear rigid*. This theoretical assumption is

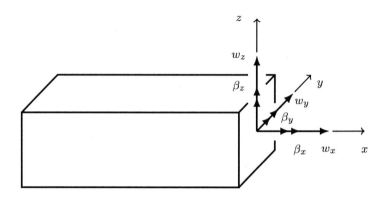

Figure 4.2: The 6 DOF at a free end of a beam and its axes

essential for obtaining handy models, and it means that stretching, torque and bending in y- and z-direction are independent from another. In order to get a better understanding of this concept, consider a homogeneous beam that is clamped at one end and free at the other[1]. If the beam is shear rigid, it does neither stretch under the impact of lateral forces, nor bend under the impact of normal forces. Additionally, all points in a y-z-cross-section are shifted equally and only in y-direction under the impact of forces in y-direction. The same holds for forces in z-direction. In the following, a mechanical system is called *beam-like*, if it can be divided into homogeneous, shear rigid beams. The definitions up next are necessary to formulate a beam element's stiffness and later to define the potential and kinetic energy of beam elements.

Definition 4.1

Let $a \in \mathbb{R}^+$ be the cross-section area of a solid beam, and $F \in \mathbb{R}$ be an external force, acting in normal direction of the beam. Then, the occurring

---

[1]Such a structure is also known as a *cantilever*.

internal force σ, given by the ratio

$$\sigma = \frac{F}{a}, \tag{4.2}$$

is called *strain*. Further, let $l \in \mathbb{R}^+$ be the length of a solid beam, and $\Delta l \in \mathbb{R}$ be a change in length, resulting from an external normal force. Then, the ratio

$$\varepsilon = \frac{\Delta l}{l} \tag{4.3}$$

is called *stress*. Let $\sigma, \varepsilon \in \mathbb{R}$ be the strain and stress of a solid beam, then the materials *elastic modulus* E is given by the ratio

$$E = \frac{\sigma}{\varepsilon}. \tag{4.4}$$

Remark 4.1

Equation (4.4), relating strain and stress of a body is known under the name of *Hooke's law*. The elastic modulus is also referred to as *Young's modulus* in literature and its dimension is force per area. This material parameter is usually determined with a tensile test, see [SGH02].

Another relevant material property is the *shear modulus*, usually denoted by G and given by

$$G := \frac{E}{2(1 + \nu)},$$

where E is the materials elastic modulus and ν is *Poisson's ratio*. This material parameter relates changes in perimeter of a beam due to stress. The restriction to shear rigid beams implies that Poisson's ratio is zero, and

$$G = \frac{E}{2}, \tag{4.5}$$

cf. [SGH02].

The cross-section of a beam element influences its stiffness also through another quantity. The following definition can be found in [SGH02] and [YB01], respectively.

Definition 4.2

Let l_x, l_y and l_z be the dimensions of a beam element with rectangular cross-section in x-, y- and z-direction, respectively. Then, the *polar moments of inertia* are given by

$$I_T := l_x l_y^3 \left(\frac{1}{3} - 0.21 \frac{l_y}{l_x} \left(1 - \frac{l_y^4}{12 l_x^4} \right) \right),$$

$$I_x := \frac{1}{12} l_y^3 l_x,$$

$$I_y := \frac{1}{12} l_x^3 l_y.$$

For circular cross-section areas with radius r, the polar moments of inertia around the x-, y- and z-axis, respectively, are given by

$$I_T := \frac{\pi r^4}{2}$$

$$I := \frac{1}{2} I_T.$$

Whenever the cross-section is square, the polar moment of inertia is also denoted by I. The approximative formula for the torsion constant I_T for elements with rectangular cross-section area was stated by Young and Budynas in [YB01]. There, the error of approximation is said to be not greater than 4%.

Equation 3.2 at the beginning of section 3.1 gave a hint to what follows. The main reason for the two definitions above is to describe a beam's ability to resist static loads. It is called stiffness, and it is divided into the abilities to resist strain, torque and bending. Under the assumptions of definition 4.1 and 4.2, a beam element's *strain stiffness* D is given by

$$D = aE, \tag{4.6}$$

where E is the materials elastic modulus and a is its cross-section area. The *torsional stiffness* T is given by

$$T = E I_T, \tag{4.7}$$

where the constant I_T has to be chosen according to the beam's cross-section, and a beam's *bending stiffness* B is given by

$$B = EI, \tag{4.8}$$

for circular and square cross-sections, regarding the adequate formula for I. Otherwise denoted $B_y = EI_y$ and $B_z = E_z$, respectively. All of these definitions are well established and can be found e.g. in [SGH02]

Definition 4.3

Let μ denote the total mass of a homogeneous beam with rectangular cross-section, and let l_x, l_y and l_z be its dimensions in x-, y- and z-direction, respectively. Then, the *moments of inertia* around the x-, y- and z-axis, respectively, are given by

$$\mu_x := \frac{1}{12}\mu\left(l_y^2 + l_z^2\right),$$

$$\mu_y := \frac{1}{12}\mu\left(l_x^2 + l_z^2\right),$$

$$\mu_z := \frac{1}{12}\mu\left(l_x^2 + l_y^2\right).$$

Let r be the radius of a homogeneous beam with circular cross-section, the moments of inertia around the x-, y- and z-axis, respectively, are given by

$$m_{x,y} := \frac{1}{4}mr^2 + \frac{1}{12}ml_z^2,$$

$$m_z := \frac{1}{2}mr^2.$$

Although the mathematical formulas are unequal, they obtain the same name since they represent the same feature of a physical body. In the case of square or circular cross-section, the moments of inertia around the y- and z axis are identical. Other than with a weighing scale, the total mass of an element can also be computed from the material's density times the element's volume, see [Ger01]. Either way is difficult to realize for complex mechanical systems that may not be decomposed.

Without an attempt of derivation[1], this subsection closes with the linear partial differential equation (PDE) that is used to model motions of shear rigid continuous beams, i.e.

$$\frac{\partial^2}{\partial x^2}\left(B(x)\frac{\partial^2}{\partial x^2}w(x,t)\right) + \mu(x)\frac{\partial^2}{\partial t^2}w(x,t) = p(x,t), \qquad (4.9)$$

with some boundary conditions, cf. [GK89b]. Here, the variable x gives the spatial localization and t stands for time. The functionals B, μ and p are finitely supported, measurable and real valued, representing the continuously distributed stiffness, mass and force, acting on the system, respectively. The solution w is again the displacement of a point x at time t. Notice that equation (4.9), also describes an ideal system, in which all energy is preserved.

4.1.2 Finite elements and spatial discretization

Within this subsection, a brief introduction to the FEM and a statement on its convergence are given. In the following subsection, these results will be used to show how real mechanical systems with continuously distributed mass and stiffness properties can be modeled by a system of second order ODE. The FEM will be used to spatially discretize the structure and its continuously distributed material properties. As the name indicates, the structure is split into a finite number of elements, each supposed to have equally distributed, i.e. constant mass and stiffness values which admit easier analyses. The dynamical behavior of one element is described by so-called ansatz functions. Gasch and Knothe suggest 6 different polynomials as ansatz functions for beam-like structures, see [GK89b].

Definition 4.4

Let $l \in \mathbb{R}^+$ and define the *ansatz functions* $f_1, f_2, f_3, f_4, g_1, g_2 : [0, l] \to \mathbb{R}$

---

[1]Chapter 6 contains a few words on this

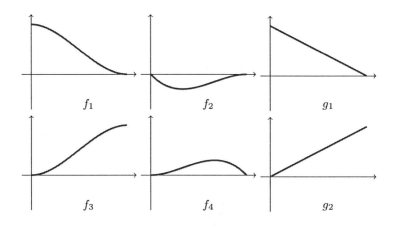

Figure 4.3: The ansatz functions, describing displacements of a beam-element

by

$$f_1(\xi) = 2\xi^3 - 3\xi^2 + 1,$$
$$f_2(\xi) = -l(-\xi + 1)^2 \xi,$$
$$f_3(\xi) = -2\xi^3 + 3\xi^2,$$
$$f_4(\xi) = l\xi^2(-\xi + 1),$$
$$g_1(\xi) = 1 - \xi$$

and

$$g_2(\xi) = \xi, \quad \text{with} \quad \xi := \frac{x}{l}.$$

The term *ansatz function* is used more generally in literature related to finite elements [ZT94], as it may refer to almost any function. Nevertheless, whenever it is used in this thesis, it refers to functions from the definition above. These functions are chosen such that they may describe changes in position, slope and torque at both ends of a beam element. Figure 4.3 shows these ansatz functions. Although it seems to be an engineering convention to denote the dimension longitudinal to the beam by z and the other two by x and y, respectively, the position longitudinal to the beam is denoted by x, here, as it is indicated in figure 4.2.

The functions $f_1, \ldots f_4$ are used to approximate the lateral motions of one element. Thus, sufficiently small displacements of the 4 lateral DOF at both ends of a beam element can be described with linear combinations of these ansatz function, cf. [GK89b]. Defining

$$f_l := [f_1, \ldots, f_4], \tag{4.10}$$

states of lateral displacement can be represented by

$$u_l f_l(\xi),$$

where $u_l \in \mathbb{R}^4$ contains the shifting and bending at both ends of the beam element, with respect to the ordering of the ansatz functions in f_l. The ansatz functions g_1 and g_2 model stretching and twisting motions. Hence, a representation for normal displacements is given by $u_n g(\xi)$ with

$$g := [g_1, g_2], \tag{4.11}$$

where $u_n \in \mathbb{R}^2$ contains the stretching at both ends of the beam element, with respect to the ordering of the ansatz functions in g. Accordingly, $u_t g(\xi)$ may represent torsional displacements.

The assumption of a shear rigid beam yields that stretching, torques and bending in y- and z-direction are independent from another, which implies that the solution of (4.9) can also be treated separately for these four cases. In the following, approximating the solution of the PDE for lateral motions in one direction, without loss of generality in y-direction, is carried out in more detail, since it reveals the main ideas behind the discretization process.

Therefore, consider a beam-like structure of total length l, theoretically divided into $n \in \mathbb{N}$ homogeneous sub-systems of lengths l_i, such that $\sum_i^n l_i = l$. For $i = 1, \ldots, n$, define $p_i : \mathbb{R}_0^+ \times [0, l_i] \to \mathbb{R}_0^+ \times [0, l_i]$, such that $p_y \equiv \sum p_i$ contains all lateral displacements in y-direction. The resulting dynamical behavior $w_i(x, t)$ of each sub-system is a solution of (4.9) with constant B_i and μ_i, for $i = 1, \ldots, n$. Thus, it holds for all at the endpoints of the integral vanishing functions $\delta x \in C^2(\mathbb{R}, \mathbb{R})$

$$\int_0^{l_i} \delta w_i(x) B_i w_i''''(x, t) + \delta w_i(x) \mu_i \ddot{w}_i(x, t) dx = \int_0^{l_i} \delta w_i(x) p_i(x, t) dx$$

$$\implies \quad \sum_{i=1}^n B_i \int_0^{l_i} \delta w_i''(x) w_i''(x, t) dx$$

$$= \sum_{i=1}^n \int_0^{l_i} \delta w_i(x) p_i(x, t) - \mu_i \delta w_i(x) \ddot{w}_i(x, t) dx, \qquad (4.12)$$

where the first summand of the first equation's left hand side is integrated by parts twice. Equation (4.12) is called *principle of virtual displacements* in the according engineering literature, see [GK89b, KW99, HK04, GHS03].

With a separation ansatz, using the space-dependent f as in (4.10) and a time-dependent displacement function $u_i : \mathbb{R}_0^+ \to \mathbb{R}^4$, lateral motions at both ends of each sub-system can be described by the product $f(x)^T u_i(t)$. By further dividing the sub-system into sub-sub-systems and so on, sufficiently small lateral motions can be described by a concatenation of ansatz functions times displacement functions, as

$$w_i(x, t) \approx f(x)^T u_i(t). \qquad (4.13)$$

Pursuing this idea for the arbitrary functions δw_i, it may also be assumed that

$$\delta w_i(x) \approx \delta u_i^T f(x), \qquad (4.14)$$

with a vector $\delta u_i \in \mathbb{R}^4$. For matters of convenience, the number of sub-systems resulting from refining the structure is also denoted by n and the lengths of each sub-systems are also denoted by l_i, such that $\sum_i^n l_i = l$. It was shown by Ivo Babuška in [BS82] that, by increasing the number of elements while reducing their length accordingly, the approximated dynamical behavior converges towards the dynamical behavior of the continuous system in some norm[1].

4.1.3 Modeling with finite elements

Based on the assumptions made on shear rigid, beam-like structures and the theoretical background of the FEM, the general procedure of discretizing these kinds of mechanical systems is explained next. The procedure is presented according to [GK89b]. It starts with spatially discretizing each term in the POVD for a theoretically isolated element. This results in two matrices representing the isolated element's stiffness and mass attributes and a vector that represents the forces that act on that very element. After repeating this for every element, these element-wise matrices are adjusted by basic transformations and summed up to two matrices, representing the system's stiffness and mass.

From equations (4.13) and (4.14) it follows that for all $i = 1, \ldots, n$

$$B_i \int_0^{l_i} \delta w_i''(x) w_i''(x,t)dx \approx B_i \int_0^{l_i} (\delta u_i^T f(x))(f''(x)^T u_i(t))dx$$

$$= \delta u_i^T B_i \int_0^{l_i} (f''(x)f''(x)^T)dx u_i(t), \qquad (4.15)$$

as well as

$$\mu_i \int_0^{l_i} \delta w_i(x) \ddot{w}_i(x,t)dx \approx \delta u_i^T \mu_i \int_0^{l_i} (f(x)f(x)^T)dx \ddot{u}_i(t). \qquad (4.16)$$

Further, with a linear approximation of the external force

$$p_i(x,t) \approx p_i(0,t)\left(1 - \frac{x}{l_i}\right) + p_i(l_i,t)\frac{x}{l_i}$$

---

[1] Just to mention that results on convergence exist

and $\xi = \frac{x}{l_i}$, it holds

$$\int_0^{l_i} \delta w_i(x) p_i(x,t) dx \approx \int_0^{l_i} \delta u_i^T f(p_i(0,t)(1-\xi) + p_i(l_i,t)\xi) dx$$

$$= \delta u_i^T \left(p_i(0,t) \int_0^{l_i} f(1-\xi) dx + p_i(l_i,t) \int_0^{l_i} f(\xi) dx \right). \quad (4.17)$$

The second derivatives of the ansatz functions are obtained by applying the chain rule, and it holds

$$f_1''(\xi) = \frac{1}{l_i^2}(12\xi - 6),$$

$$f_2''(\xi) = \frac{1}{l_i}(-6\xi + 4),$$

$$f_3''(\xi) = \frac{1}{l_i^2}(-12\xi + 6),$$

$$f_4''(\xi) = \frac{1}{l_i}(-6\xi + 2).$$

Thus, the values of the dyadic product in (4.15), i.e.

$$f'' f''^T = \begin{bmatrix} f_1'' f_1'' & \cdots & f_1'' f_4'' \\ \vdots & \ddots & \vdots \\ f_4'' f_1'' & \cdots & f_4'' f_4'' \end{bmatrix},$$

can be computed, such that

$$B_i \int_0^{l_i} (f''(\xi) f''(\xi)^T) dx = \frac{B_i}{l_i^3} \begin{bmatrix} 12 & -6l_i & -12 & -6l_i \\ -6l_i & 4l_i^2 & 6l_i & 2l_i^2 \\ -12 & 6l_i & 12 & 6l_i \\ -6l_i & 2l_i^2 & 6l_i & 4l_i^2 \end{bmatrix}. \quad (4.18)$$

The matrix in (4.18) quantitatively describes one element's ability to hold lateral loads, and it will be denoted by S_B. Due to the restriction to homogeneous beams, the bending stiffness B_i changes only under variations of

47

the cross section area. In the case of circular, or square cross sections, this matrix is identical for bending towards y- and z- direction. Without loss of generality, the case of 2 different bending stiffnesses is carried out in the following, giving two matrices (4.18) called S_{B_y} and S_{B_z}.

On an analogous way, two matrices S_D and S_T are determined, using the separation ansatz (4.13) and (4.14), with ansatz functions g as in (4.11) instead of f, displacement functions $u_D, u_T : \mathbb{R}_0^+ \to \mathbb{R}^2$ and displacement vectors $\delta u_D, \delta u_T \in \mathbb{R}^2$, respectively. It holds

$$S_D = \frac{D}{l_i} \begin{bmatrix} 1 & -1 \\ -1 & 1 \end{bmatrix}, \quad S_T = \frac{B_x}{l_i} \begin{bmatrix} 1 & -1 \\ -1 & 1 \end{bmatrix}. \qquad (4.19)$$

The matrices S_D and S_T represent one element's ability to resist normal and torsional loads, respectively.

All 4 matrices above enter the *element stiffness matrix* $\hat{S}_i$ linearly independent due to the assumption of shear rigidity.

$$\hat{S}_i := \begin{bmatrix} S_D & & & \\ & S_T & & \\ & & S_{B_y} & \\ & & & S_{B_z} \end{bmatrix}. \qquad (4.20)$$

The construction of the *element mass matrix* $\hat{M}_i$ takes an according form, i.e.

$$\hat{M}_i := \begin{bmatrix} M_D & & & \\ & M_T & & \\ & & M_{B_y} & \\ & & & M_{B_z} \end{bmatrix}, \qquad (4.21)$$

using (4.16). The separation ansatz (4.13) and (4.14) are applied also for the ansatz functions g as in (4.11). The total mass and all moments of inertia are denoted according to definition 4.3. Thus, an element's inertia attributes relative to normal and torsional forces, respectively are described by M_D and M_T, which are given by

$$M_D = \frac{\mu l_i}{6} \begin{bmatrix} 2 & 1 \\ 1 & 2 \end{bmatrix}, \quad M_T = \frac{\mu_x l_i}{6} \begin{bmatrix} 2 & 1 \\ 1 & 2 \end{bmatrix}. \tag{4.22}$$

Inertia attributes relative to lateral forces are given in the form

$$M_{B_y} := \frac{\mu l_i}{420} \begin{bmatrix} 156 & -22l_i & 54 & 13l_i \\ -22l_i & 4l_i^2 & -13l_i & -3l_i^2 \\ 54 & -13l_i & 156 & 22l_i \\ 13l_i & -3l_i^2 & 22l_i & 4l_i^2 \end{bmatrix}$$

$$+ \frac{\mu_y}{30l_i} \begin{bmatrix} 36 & -3l_i & -36 & -3l_i \\ -3l_i & 4l_i^2 & 3l_i & -l_i^2 \\ -36 & 3l_i & 36 & 3l_i \\ -3l_i & -l_i^2 & 3l_i & 4l_i^2 \end{bmatrix}, \tag{4.23}$$

and an according matrix M_{B_z} since the element's moments of inertia around the y- and z-axis may be unequal. In the same manner, the element load function, see (4.17), is separated into its normal component $p_{i,n}$, its torsional component $p_{i,mx}$ and its lateral components p_{i,l_y} and p_{i,l_z}. This gives

$$\hat{P}_i^T(t) := \left[P_D(t), P_T(t), P_{B_y}(t), P_{B_z}(t) \right], \tag{4.24}$$

which consists of

$$P_D(t) := \frac{l_i}{6} \left(p_{i,n}(0,t) \begin{bmatrix} 2 \\ 1 \end{bmatrix} + p_{i,n}(l_i,t) \begin{bmatrix} 1 \\ 2 \end{bmatrix} \right),$$

$$P_T(t) := \frac{l_i}{6} \left(p_{i,mx}(0,t) \begin{bmatrix} 2 \\ 1 \end{bmatrix} + p_{i,t}(l_i,mx) \begin{bmatrix} 1 \\ 2 \end{bmatrix} \right),$$

$$P_{B_y}(t) := \frac{l_i}{10} \left(p_{i,l_y}(0,t) \begin{bmatrix} 7/2 \\ -l_i/2 \\ 3/2 \\ l_i/3 \end{bmatrix} + p_{i,l_y}(l_i,t) \begin{bmatrix} 3/2 \\ -l_i/3 \\ 7/2 \\ l_i/2 \end{bmatrix} \right)$$

$$+ \frac{1}{2} \left(p_{i,mz}(0,t) \begin{bmatrix} -1 \\ -l_i/6 \\ 1 \\ l_i/6 \end{bmatrix} + p_{i,mz}(l_i,t) \begin{bmatrix} -1 \\ l_i/6 \\ 1 \\ -l_i/6 \end{bmatrix} \right),$$

and an analogue form for P_{B_z}. The mx denotes the applied torque, which results also in a moment around x. In the same sense, the quantity mz denotes a moment around z. As it was described in the beginning of subsection 4.1.1, a moment load mz forces the beam to arch towards y.

Constructing the element stiffness and mass matrix according to (4.20) and (4.21), as well as the element load vector (4.24) admits the approximation of each summand in (4.12) by an equation of the form

$$\delta u^T \hat{M} \ddot{u}(t) + \delta u^T \hat{S} u(t) = \delta u^T \hat{P}, \qquad (4.25)$$

where $\delta u \in \mathbb{R}^{12}$ and $u : \mathbb{R}_0^+ \to \mathbb{R}^{12}$ are concatenations of the displacement vectors and functions, respectively, related to the according separation ansatz for lateral, normal and torsional deflections.

As a next step, these matrices may be re-ordered in the way that the entries in u give the displacement in x-, y- and z-direction, the torsion and the slope in y- and z-direction at the upper end of the element and then the displacements, torsion and slopes at the lower end of the element in that order. Conventionally, as described in [GK89b], the sign of the slopes in y- and z-direction are changed, too. This process is realized by the multiplication

$$M_i = T^T \hat{M} T, \quad S_i = T^T \hat{S} T,$$

of the element stiffness and mass matrix with the transformation matrix

$$
T = \begin{bmatrix}
1 & 0 & 0 & 0 & 0 & 0 & 0 & 0 & 0 & 0 & 0 & 0 \\
0 & 0 & 0 & 0 & 0 & 0 & 1 & 0 & 0 & 0 & 0 & 0 \\
0 & 0 & 0 & 1 & 0 & 0 & 0 & 0 & 0 & 0 & 0 & 0 \\
0 & 0 & 0 & 0 & 0 & 0 & 0 & 0 & 0 & 1 & 0 & 0 \\
0 & 0 & 1 & 0 & 0 & 0 & 0 & 0 & 0 & 0 & 0 & 0 \\
0 & 0 & 0 & 0 & 1 & 0 & 0 & 0 & 0 & 0 & 0 & 0 \\
0 & 0 & 0 & 0 & 0 & 0 & 0 & 0 & 1 & 0 & 0 & 0 \\
0 & 0 & 0 & 0 & 0 & 0 & 0 & 0 & 0 & 0 & 1 & 0 \\
0 & 1 & 0 & 0 & 0 & 0 & 0 & 0 & 0 & 0 & 0 & 0 \\
0 & 0 & 0 & 0 & 0 & -1 & 0 & 0 & 0 & 0 & 0 & 0 \\
0 & 0 & 0 & 0 & 0 & 0 & 0 & 1 & 0 & 0 & 0 & 0 \\
0 & 0 & 0 & 0 & 0 & 0 & 0 & 0 & 0 & 0 & 0 & -1
\end{bmatrix}. \tag{4.26}
$$

This order is also used for the final displacement vector and matrices that represent the complete mechanical system. Therefore, further transformation matrices T_i are introduced. These matrices are constructed to fulfill two purposes. First, the transformation writes the element related matrices into a matrix, the size of the final system, and second, all bearings, supports and clamps of the mechanical system are realized by nullifying the according DOF. The final system matrices are obtained by

$$
S := \sum_i T_i^T S_i T_i \quad \text{and} \quad M := \sum_i T_i^T M_i T_i.
$$

A more descriptive explanation of the transformation matrices's form is given in the next subsection. Since the displacement vector $\delta u \in \mathbb{R}^{12}$ is arbitrary

$$
M\ddot{u}(t) + Su(t) = P(t) \tag{4.27}
$$

can be used as a model for simulations of the dynamical behavior of a beam-like structure. To this end, let f be an appropriately ordered concatenation of the ansatz functions, defined in (4.10) and (4.11). Then, with a solution

$u_0(t)$ of (4.27) the vibrational response of the objective structure to a load P can be visualized by

$$u_0(\cdot)f(\cdot) : \mathbb{R}_0^+ \times [0, l] \longrightarrow \mathbb{R} \times [0, l],$$

where $l \in \mathbb{R}^+$ denotes the total length of the objective.

The procedure, described above, demands several remarks.

Remark 4.2   1. The dimension of the matrices M and S corresponds to the number of modeled DOF of the structure.

2. In the case of beam-like structures as described above, the resulting matrices M and S will always be real and symmetric.

3. The order of the ansatz functions in (4.10), (4.11) and the order of the matrices in (4.20),(4.21) is arbitrary, but dictates the face of the final model. Here, it is chosen according to [GK89b].

4. The test functions, used to solve the minimization problem in equation (4.12) are known as *virtual displacements* among engineers. Besides being virtual and arbitrary, they are understood as differentially small displacements that have to be geometrically possible and immediate, i.e. not depending on time. Equation (4.12) itself is called the *principle of virtual displacements* (POVD) in the engineering science, and integrating by parts twice reveals that it is equivalent to the PDE (4.9).

4.1.4 Example

Within this subsection, the FEM based approach to model beam-like structures is exemplified. The modeled device is a *DMG Ultrasonic 20 linear* milling machine, setup at the Labor für Mikrozerspanung (LfM), Bremen. The machine itself is displayed in figure 4.1, and the dimensions of the part which is modeled are displayed in figure 4.4. Further, the following explanation may help to understand, how computing FEM models for beam-like

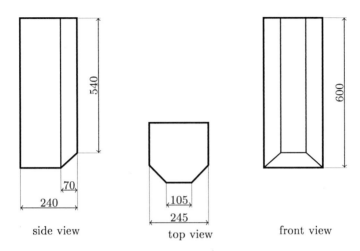

side view   top view   front view

Figure 4.4: Dimensions of the objective machine tool's top mount

structures from given geometrical and material properties can automated. An according algorithm was implemented in Matlab.

For a proper application of the FEM routine, presented above, some simplifying assumptions to the structure have to be made. Therefore, the trapezoidal front-part of the upper mount is rectified and combined with the rest of the upper mount into one cuboid. This is done, preserving its original volume. The truncated cone, located at the lower end of the middle shaft is theoretically transformed into a cylinder of equal volume. All simplifications and the new dimensions are displayed in figures 4.5 and 4.6. The tool holder itself needs no simplification. Its dimensions are presented in figure 4.7, together with a quantitative sketch of the whole simplified structure. All dimensions are given in millimeters.

Because this section is meant as an additional explanation of how to apply the FEM procedure, there is no attempt for theoretical justification or precision of the later model. The structure is divided into elements such that

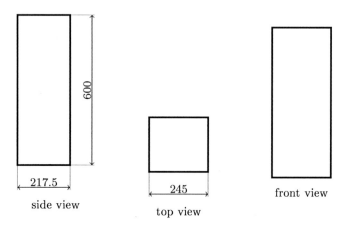

Figure 4.5: Dimensions of the simplified machine tool's top mount

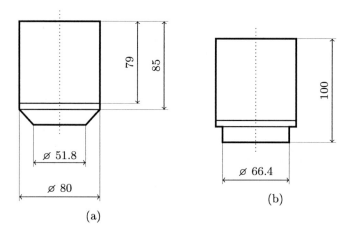

Figure 4.6: Dimensions of the machine tool's shaft in (a) and a simplification in (b)

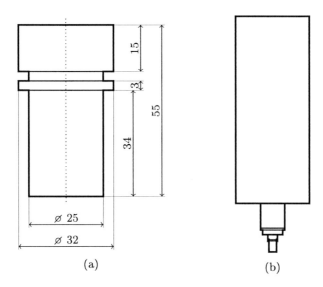

Figure 4.7: Dimensions of the tool holder in (a) and quantitative sketch of
the simplified structure in (b)

within one element no changes of the cross section or material occur, as indicated in figures 4.6 and 4.7. Since the structure could not be decomposed,
changes of the material were decided by sight. As a result, 11 elements are
obtained. These are the top mount, assumed to be clamped at the upper
end, divided into 2 elements, the lower mount, divided into 4 elements and
the tool-holder, divided into 5 elements. Thus, the machine will be modeled
by two 66 × 66 matrices, i.e. 66 DOF. A finer mesh may result in higher
accuracy of the model, although there are no theoretical results on how to
choose the mesh.

The material of the lower shaft is known to be tungsten carbide, which has
an elastic modulus of $E_T = 650\text{kN}/\text{mm}^2$, a density of $d_T = 14.5 \cdot 10^{-3}\text{g}/\text{mm}^3$
and a shear modulus of $G_T = 180\text{kN}/\text{mm}^2$, cf. [She07]. The rest of the
structures material properties are not known exactly and have to be esti-

mated. As an initial guess and for the purpose of demonstration, all other materials are assumed to be solid carbon steel with an elastic modulus of $E_S = 210\text{kN/mm}^2$, a density of $d_S = 7.85 \cdot 10^{-3}\text{g/mm}^3$ and a shear modulus of $G_S = 79\text{kN/mm}^2$, cf. [Moe08]. In order to cope with some of the structure's components not being completely solid, the elastic modulus, density and shear modulus of the materials in the example model have been adjusted. Parameter identification methods may by applied in order to improve the model, see section 5.2. Starting from the upper end, the first

Figure 4.8: Schematic presentation of element-wise matrix

element is the top mount. From definitions 4.2 and 4.3, the elements bending stiffness and mass attributes can be determined, and the element's 12×12 stiffness and mass matrix are determined, following (4.20) and (4.21). These two matrices are multiplied from right and left with matrix T and T^T, respectively, as described in (4.26). Before this transformation, the element's stiffness and mass matrices take the form displayed in figure 4.8. There, the ×-symbol indicates a positive and the -× a negative scalar, and empty entries represent zero entries. After the first transform, they take the form displayed in figure 4.9, with the same notation as in figure 4.8.

$$
\begin{bmatrix}
\times & & & & & & -\times & & & & \\
& \times & & & & \times & -\times & & & \times \\
& & \times & -\times & & & -\times & & -\times & \\
& & & \times & & & & -\times & & \\
& & -\times & \times & & & \times & & \times & \\
& \times & & & & \times & -\times & & & \times \\
-\times & & & & \times & & & & & \\
& -\times & & & -\times & \times & & & & -\times \\
& -\times & \times & & & & \times & & \times & \\
& & -\times & & & & & \times & & \\
& -\times & \times & & & & \times & & \times & \\
& \times & & & & \times & -\times & & & \times \\
\end{bmatrix}
$$

Figure 4.9: Schematic   presentation   of   element-wise   matrix   after   1st transformation

Now, the first 6 entries in the according element displacement vector, see (4.25), represent the 6 DOF from the upper end of the top mount. Since this end of the structure is assumed to be clamped, it can not move in these DOF! The according entries of the element stiffness and mass matrices are therefore set to zero with the second transformation. Additionally, the matrices are transformed to match with the final size of the system. In the special case discussed here, the second transformation matrix for the first element, denoted by T_1, is a 12×66 sparse matrix with a downward diagonal of ones, starting from the seventh entry of the first column and zeros otherwise. The transformation is made by a right and left multiplication with T_1 and T_1^T, respectively, and it takes the form displayed in figure 4.9, where again the ×-symbol indicates a positive and the -× a negative scalar, but zero entries are explicitly indicated.

$$
\begin{bmatrix}
\times & 0 & 0 & 0 & 0 & 0 & 0 & \cdots & 0 \\
0 & \times & 0 & 0 & 0 & -\times & 0 & & \vdots \\
0 & 0 & \times & 0 & \times & 0 & 0 & & \\
0 & 0 & 0 & \times & 0 & 0 & 0 & & \\
0 & 0 & -\times & 0 & \times & 0 & 0 & & \\
0 & -\times & 0 & 0 & 0 & \times & 0 & & \\
0 & 0 & 0 & 0 & 0 & 0 & 0 & & \\
\vdots & & & & & & & \ddots & \vdots \\
0 & \cdots & & & & & & \cdots & 0
\end{bmatrix}
.
$$

Figure 4.10: Schematic presentation of element-wise matrix after 2nd transformation

The material parameters, stiffness and mass matrices for the other elements are determined in the same manner. While first transformation is performed identically for every element, the second transformation matrix changes for every element. Each matrix is a 12×66 zero-matrix with a downward diagonal of ones, but the starting point of the diagonal shifts. For the second element, the diagonal starts at the first entry of the seventh column and goes all the way down to the 12th element of the 18the column. The third starts from the first entry of the 13th column and so on! The resulting structure of both, the system's stiffness and mass matrix is shown in figure 4.11. The dots substitute a continuation of the alternating long- and short-antidiagonal pattern. Zero entries and the indication of each entry's sign are omitted.

Figure 4.11: Schematic and abbreviated representation of a system matrix

The sample setup has been analyzed numerically, and a simulation[1] of its dynamical behavior is shown in figure 4.12. The figure shows the vibrational response at 4 of the 11 nodes of the model to an initial displacement of the free end in y-direction. Beginning from the top, the first graph belongs to the lowest point of the lower mount, and the bottom graph describes the motion of the free end of the structure. The two graphs in between describe motions of the tool holder at chosen positions. The simulated time is 0.2ms.

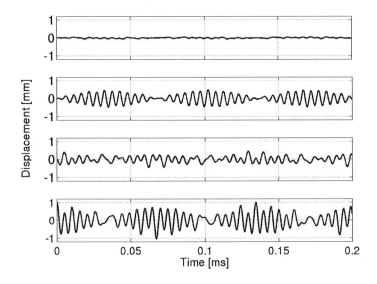

Figure 4.12: Simulated dynamical behavior in 4 selected DOF

---

[1]The according ODE was solved with the Matlab routine ode23t.

4.2 Solving oscillator equations

In the previous section, it was shown how to model beam-like structures by a damped mass-spring system with arbitrary s-dimensional real matrices M and S as seen in (4.27). The next step in simulating dynamical behavior of mechanical structures is to solve the ODE, derived in the previous section, for a given set of material and structural parameters as well as a given excitation. In the following, a brief recall of basic ODE theory and some advanced results in this field are presented. The presentation relies on the textbooks "Gewöhnliche Differentialgleichungen" by Dirk Walter and "Mathematical Control Theory" by Eduardo D. Sontag, see [Wal00, Son98].

Taking mass-spring systems as they are presented in (4.1), the existence of a unique solution is guaranteed for a wide range of right hand sides. Since the Fourier-transform is a bijection on L^2, for all $f \in L^2$, a unique solution of (4.1) may be obtained by

$$
\begin{aligned}
& M\ddot{u}(t) + Su(t) = f(t) \\
\Longleftrightarrow \quad & (-\omega^2 M + S)\mathscr{F}u(\omega) = \mathscr{F}f(\omega) \\
\Longleftrightarrow \quad & u(t) = \mathscr{F}^{-1}(-\omega^2 M + S)^{-1}\mathscr{F}f(\omega)(t).
\end{aligned}
\tag{4.28}
$$

Especially if a system is excited by a harmonic load only, vibrational responses can be computed analytically with this Fourier-transform approach.

In general, any system of second order ODE can be transformed into a larger system of first order, see [Wal00, Ama90, HrW87]. With

$$
x(t) := \left[\begin{array}{c} u(t) \\ \frac{d}{dt}u(t) \end{array} \right],
\tag{4.29}
$$

$$
A := \left[\begin{array}{cc} 0 & \mathbb{1}_s \\ -M^{-1}S & 0 \end{array} \right],
\tag{4.30}
$$

and

$$F(t) := \begin{bmatrix} 0 \\ M^{-1}f(t) \end{bmatrix}, \tag{4.31}$$

equation (4.1) is equivalent to

$$\dot{x}(t) = Ax(t) + F(t). \tag{4.32}$$

The equation is called *homogeneous* if no exterior forces are applied, i.e. $F \equiv 0$. Finding a solution of (4.32) for given time t_0 and $x(t_0) = x_0$ is called an *initial value problem* (IVP). A theorem, stating prerequisites for the existence and uniqueness of solutions of IVP is known under the name of Picard-Lindelöf. An advanced version of the well-known Picard-Lindelöf theorem with less restrictive prerequisites is known under the name of *Carathéodory's theorem*. In the context of mechanical systems, the advantage of Carathéodory's theorem in comparison with Picard-Lindelöf is that it admits a wider rage of load transmissions. The theorem itself can be found e.g. in [Wal00] and [Son98], while the latter also contains a proof.

Theorem 4.3 (Carathéodory)
Let $I \subset \mathbb{R}$ be an interval, $\Omega \subset \mathbb{R}^d$ be open and $f : I \times \Omega \to \mathbb{R}^d$. Further, let $f(t, \cdot)$ be continuous for every $t \in I$ and $f(\cdot, x)$ be measurable for every $x \in \Omega$ and fulfill the following conditions:

1. *For every fixed x_0 there exists a locally integrable function $\alpha : I \to \mathbb{R}^+$, such that*

$$\|f(t, x_0)\| \leq \alpha(t), \quad a.e.$$

2. *For every $x_0 \in \Omega$ there exists an $\varepsilon > 0$, such that the open ball $B_\varepsilon(x_0)$ is contained in Ω and a locally integrable function $\beta : I \to \mathbb{R}^+$ exists, such that*

$$\|f(t, x) - f(t, \tilde{x})\| \leq \beta(t)\|x - \tilde{x}\|$$

for all $t \in I$ and $x, \tilde{x} \in B_\varepsilon(x_0)$

Then, the IVP (4.32) has a unique solution in the sense of Carathéodory.

This means that for each pair $(t_0, x_0) \in I \times \Omega$ a proper subinterval $J \subset I$, relatively open to I, and a solution $x : J \to \mathbb{R}^d$ of (4.32) exists with the following property. If $y : I \supset J' \to \mathbb{R}^d$ is another solution of (4.32), then necessarily

$$J' \subset J \text{ and } y = x \text{ on } J'.$$

The solution x is also called the *maximal solution* of the IVP (4.32).

The above theorem guarantees the existence of a unique solution of IVP, describing the dynamical behavior of mass-spring systems with load transmissions of impulse-type. With a constant system matrix A, variation of constants yields that all solutions x take the form

$$x(t) = e^{(t-t_0)A} x_0 + \int_{t_0}^t e^{(t-s)A} F(s) ds, \qquad (4.33)$$

where the pair (t_0, x_0) gives the initial value. Like the theorem and its proof, this statement can be found in [Son98]. The special case, where $t_0 = 0$ and $x(0) = 0$ is considered in the following and thus it holds

$$x(t) = \int_0^t e^{(t-s)A} F(s) ds, \qquad (4.34)$$

which is a special case of a *Volterra integral equation of the first kind.* In general, these equations are noted in the form

$$x(t) = \int_0^t k(s,t) y(s) ds, \qquad (4.35)$$

where k is called the *integral kernel*, cf. [Lue95, Kre99].

Solutions of the homogeneous system are of the form

$$x(t) = v \cdot e^{\lambda t} = \begin{bmatrix} v_1 e^{\lambda t} \\ \vdots \\ v_n e^{\lambda t} \end{bmatrix}, \qquad (4.36)$$

where λ is an eigenvalue of A, and v is the corresponding eigenvector. For the case of a damped oscillator equation, i.e. A as in (4.30), these eigenvalues can be computed, using the special structure of A.

It is known from linear algebra that for all $s \times s$ matrices X_1, X_2 and X_3 that $\det(X_1)\det(X_2) = \det(X_1 X_2)$ and

$$\det \begin{bmatrix} X_1 & X_2 \\ 0 & X_3 \end{bmatrix} = \det \begin{bmatrix} X_1 & 0 \\ X_2 & X_3 \end{bmatrix} = \det(X_1)\det(X_3),$$

cf. [Fis02]. Thus, it can be seen through the representation of $(A - \lambda \mathbb{1}_s)$

$$\begin{bmatrix} -\lambda \mathbb{1}_s & \mathbb{1}_s \\ -M^{-1}S & -\lambda \mathbb{1}_s \end{bmatrix} = \begin{bmatrix} -\lambda \mathbb{1}_s & 0 \\ -M^{-1}S & \mathbb{1}_s \end{bmatrix} \cdot \begin{bmatrix} \mathbb{1}_s & -\lambda^{-1}\mathbb{1}_s \\ 0 & -\lambda \mathbb{1}_s - \lambda^{-1}M^{-1}S \end{bmatrix}$$

that

$$\det(A - \lambda \mathbb{1}_s) = \det(-\lambda \mathbb{1}_s)\det(-\lambda \mathbb{1}_s - \lambda^{-1}M^{-1}S)$$
$$= \det(\lambda^2 \mathbb{1}_s + M^{-1}S).$$

All solutions of an inhomogeneous ODE are obtained by adding one special solution of the inhomogeneous to a solution of the homogeneous ODE. Hence, they play a special role for the mechanical system behind the equations, since they characterize its natural dynamical behavior. For the case of 1 DOF, i.e. $s = 1$, these eigenvalues coincides with the solutions of the QEP (3.6) and an extension of definition 3.2 from the previous chapter can be made.

Definition 4.5
The imaginary part of each eigenvalue of A as in (4.30) is called a *natural frequency of M and S*.

4.2.1 Explicit one-step methods

Now that the theoretical background of the ODE according to structural dynamics are set, the practical side of solving IVP numerically is the goal of this subsection. As the title reveals, the presentation is limited to explicit one-step methods (EOSM). A brief motivation of the general concept of one-step methods will lead to the first basic example, the Euler-method. The final result of this section is the introduction of the Runge-Kutta method (RKM), a well established and powerful tool for the numerical solution of differential equations. Within this and the preceding chapter, Matlab implementations of EOSM have been used to simulate vibrational responses of mechanical systems by solving the according ODE numerically. The basic information on numerical methods for solving differential equations given here can also be found in [SB92, HrW87, Ise98]. In this section, all IVP are assumed to be uniquely solvable.

Given a general IVP

$$\dot{x}(t) = f(t, x(t)), \ x(t_0) = x_0, \tag{4.37}$$

it holds approximately

$$\frac{x(t+h) - x(t)}{h} \approx f(t, x(t))$$

for $h \neq 0$ and thus,

$$x(t+h) \approx x(t) + hf(t, x(t)).$$

Using this observation, the *polygon method* of Euler estimates solutions η_k of $x(t_k)$ at equidistant points $t_k = t_0 + kh$, for $k = 1, 2, \ldots$, as follows:

$$\eta_0 := x_0,$$

and for $k = 1, 2, \ldots$:

$$\eta_{k+1} := \eta_k + hf(t_k, \eta_k).$$

It is shown, e.g. in [SB92] that the error of Euler's method decreases linearly when the step-size is reduced.

In 1895, Carl Runge published an article on the numerical solution of differential equations [Run95] in which he revised Euler's idea. Runge showed the connection of the Euler method to the trapezoidal rule for numerical integration and that their precision depends linearly on the step size. As an improvement, he suggested an algorithm strongly connected to Simpson's rule, which Martin Wilhelm Kutta refined in 1901, introducing the well-known RKM [Kut01].

A η_k of $x(t_k)$ at equidistant points $t_k = t_0 + kh$, for $k = 1, 2, \ldots$, as follows:

$$\eta_0 := x_0,$$

and for $k = 1, 2, \ldots$:

$$\eta_{k+1} := \eta_k + h\Phi(t_k, \eta_k, h).$$

where $\Phi(t, \eta, h) := \frac{1}{6}(k_1 + 2k_2 + 2k_3 + k_4)$, and

$$k_1 := f(t, x(t)),$$
$$k_2 := f(t + \frac{1}{2}h, x(t) + \frac{1}{2}hk_1),$$
$$k_3 := f(t + \frac{1}{2}h, x(t) + \frac{1}{2}hk_2),$$
$$k_4 := f(t + h, x(t) + hk_3).$$

The error of the RKM decreases with the order 4 when the step-size is reduced, cf. [SB92]. A thorough convergence analysis of one-step methods, as well as methods of higher order can also be found in [SB92].

4.3 Damping models for mechanical systems

Within this final section of the chapter, dedicated to describing structural dynamics, the process of modeling will be completed. As it was done in the previous chapter for the 1 DOF case, the last step is incorporating the dissipation of energy, or damping. Section 4.1 revealed that within the process of computing the system matrices M and S with the FEM, no kind of damping matrix is produced. It is commonly accepted among engineers that damping properties in general are rarely known and difficult to evaluate precisely, see e.g. [GK89a, TM01]. As in section 3.2, the dissipation of energy is assumed to be proportional to the system's velocity, cf. [WB06, WJ87, GK89a].

In order to approach a damping model, (4.27) is augmented with an arbitrary operator D, acting linearly on the system's velocity. This gives

$$M\ddot{u}(t) + D\dot{u}(t) + Su(t) = f(t). \tag{4.38}$$

Analogously to (4.30), the ODE can be transformed into a first order system $\frac{d}{dt}x(t) = Ax(t) + F(t)$, where

$$A := \begin{bmatrix} 0 & \mathbb{1}_s \\ -M^{-1}S & -M^{-1}D \end{bmatrix}. \tag{4.39}$$

From the abstract point of view, taken in the previous section, a proposition in Eduardo Sontag's textbook on mathematical control theory [Son98] contributes to the understanding of damped mass-spring systems.

Proposition 4.4

Let A be a real $n \times n$ matrix. Then a solution x of

$$\dot{x}(t) = Ax(t) \tag{4.40}$$

satisfies

$$x(t) \to 0 \text{ as } t \to \infty \tag{4.41}$$

if and only if all eigenvalues of A have negative real part.

A formal proof is also given in [Son98]. In the case where A results from transforming (4.38) into a first order system, D is called a *damping matrix*, if and only if all eigenvalues of A have negative real part. In this case (4.38) is also referred to as a *damped mass-spring system* in the engineering literature.

In the 1 DOF case, described by the positive integers m and s, this proposition gives a condition for the damping coefficient c, introduced in (3.13). There, the eigenvalues of the first order system are given by

$$\det(A - \lambda \mathbb{1}_2) = \det \begin{bmatrix} -\lambda & 1 \\ -\frac{s}{m} & -\frac{c}{m} - \lambda \end{bmatrix} = \lambda^2 + \lambda \frac{c}{m} + \frac{s}{m} = 0 \quad (4.42)$$

and coincide with the solutions of the aforementioned QEP (3.14). The real part of these eigenvalues is strictly negative if and only if $c > 0$, cf. [Hur95], and the eigenvalues are purely imaginary if and only if $c = 0$.

Due to the original article "Ueber die Bedingungen, unter welchen eine Gleichung nur Wurzeln mit negativen reellen Theilen besitzt" by Adolf Hurwitz from 1895, a matrix A is also called *Hurwitz* if all its eigenvalues lie on the left complex plane, see [Hur95] and [Son98]. In the multi DOF case, the criterion to check whether all eigenvalues of A have negative real part concerns scientists until now. In [Son98], several possible tests for this purpose are suggested, including the so called Hurwitz-Routh test and a method employing a Lyapunov function. A proof of the Hurwitz-Routh test can be found in [Mei95], and an introduction to Lyapunov functions can be found in the context of general control theory, e.g. [Son98, HP05].

Based on these results definition 3.5 from the previous chapter can be extended to the multi DOF case.

Definition 4.6

The imaginary part of each eigenvalue of A as in (4.39) is called an *eigenfrequency of M, D and S*.

In the case where M, D and S describe a mechanical system, each eigenfrequency corresponds to one of the modeled DOF.

4.3.1 Rayleigh damping

The so called Rayleigh damping model is recently tested and used extinctively, see e.g. [Mee08]. Here, the available information about the system, gathered in the matrices M and S, is used in the sense of a linear combination in order to incorporate energy dissipation. In [Cau60] damping of mechanical systems is modeled via the sum of $n \in \mathbb{N}$ matrices.

Definition 4.7
Let V be a real vector space, and let $M, S \in L(V, V)$. For $n \in \mathbb{N}$, and $\sigma_k \in \mathbb{R}^+$,

$$D := M \sum_{k=0}^{r} \sigma_k (M^{-1}S)^k.$$

is called *Rayleigh damping matrix*.

Meerbergen as well as Liu consider only the special case of Rayleigh damping, where $r = 1$ [LG95, Mee08]. Thus, the damping matrix D takes the form

$$D = \alpha M + \beta S, \tag{4.43}$$

with appropriately chosen coefficients $\alpha, \beta \in \mathbb{R}_0^+$, and mass and stiffness matrices M and S, respectively.

For a structure without actual damping elements, the coefficient β will vanish and the damping is called structural damping. The coefficients α and β can be included in parameter identification routines, cf. 5.2. In the case where D is the Rayleigh damping matrix of a beam-like structure, it will always be real, symmetric and positive definite.

Combining the FEM process, described in section 4.1.3 and the Rayleigh damping matrix (4.43), an operator may be defined that maps elements p from a parameter space $\mathbb{P}$ to a triple of real, symmetric and positive definite matrices by

$$\mathfrak{F}_s : \mathbb{P} \to L(\mathbb{R}^s)^3$$
$$\mathfrak{F}_s(p) := (M, D, S). \tag{4.44}$$

The maximal dimension n of $\mathbb{P}$ is determined by all possibly variable parameters, i.e. three parameters for length, density, elastic modulus and shear modulus for each element, taken into account, plus 2 parameters for the Rayleigh damping matrix. For a cantilever beam modeled with s DOF, the relation between n and s is thus given by

$$n \leq 3s + 2.$$

5 The Inverse Problem

Considering a certain problem an inverse problem, usually means that its approach from another point of view is easier or even trivial. In the case, where computing a mechanical system's reaction to a certain load transmission is considered the forward problem, a number of inverse problems arise naturally. The following seem to be of greater interest among scientists and will be discussed more or less elaborate within this chapter.

- Optimizing an existing mathematical model of the system is referred to as model updating. By means of model updating, a mathematical model of an existing structure may be improved by exploiting actual measurements. This process can be crucial to further theoretical analyses of structures. One mathematical approach to model updating will be presented in section 5.2.

- The general desire to understand the origin of an observed vibrational response is the motivation behind section 5.3. Here, unknown load transmissions to a given mechanical structure will be recovered from measurements of its vibrational response. The main results and contribution to the field of structure dynamics and vibrational analysis, presented in this thesis, are also found within this section.

- In order to obtain a prescribed dynamical behavior, either the load transmission or the structure itself may be manipulated. Thus, problems of this kind are tackled from various positions.

- One possible way of determining necessary manipulations of the load transmission is called input shaping and will be reviewed in section 5.4.

- Methods that manipulations the structure are known as node assignment. These methods are further devided into passive modification or active control methods, depending on whether masses and springs or so called activators are used. Node assignement will not be discussed in this thesis, but a comprehensive overview is given by Mottershead in [MR06].

The first section of this chapter presents an introduction to the theory behind the methods, applied to solve the different inverse problems.

5.1 Well-conditioned and well-posed problems

Solving an inverse problem does not always involve big obstacles. Simple matrix-vector multiplication

$$Ax = b,$$

in a finite dimensional vector space V, with an isomorphism $A \in L(V, V)$ and $x \in V$ may of course be called a forward problem. Thus, computing x to a given $b \in V$ is an inverse problem, but a unique solution is guaranteed, and the solution x of the inverse problem depends continuously on perturbations of b. Nevertheless, inquiring the sensitivity of perturbations of x to changes in b lead to the following definition.

Definition 5.1
Let V be a finite dimensional vector space, $A \in L(V, V)$ and $\| \cdot \|$ is a matrix norm. Then, the *condition of* A is given by

$$\mathrm{cond}(A) := \|A\| \cdot \|A^{-1}\|.$$

This is a standard definition from linear algebra or numerical analysis and can be found, e.g. in [SB92]. It gives a measure of the sensitivity of relative errors in the solution to changes in the right-hand side of a linear equation. Matrices with a condition close to 1 are considered *well-conditioned*.

This definition is utilized to make a statement about the solvability of linear systems with perturbed data.

Proposition 5.1

Let x be the unique solution of $Ax = b$, where $A \in L(V, V)$ and $b \in V$. Let A^δ and b^δ be perturbations of A and b, respectively, such that

$$q = \text{cond}(A) \frac{\|A - A^\delta\|}{\|A\|} < 1.$$

Then, there exists a unique solution to the perturbed equation

$$A^\delta x^\delta = b^\delta,$$

and it holds

$$\frac{\|x - x^\delta\|}{\|x\|} \leq \frac{\text{cond}(A)}{1 - q} \left(\frac{\|A - A^\delta\|}{\|A\|} + \frac{\|b - b^\delta\|}{\|b\|} \right).$$

A proof of this proposition can be found in [SB92].

In vector spaces of infinite dimensions, such as the function spaces C^∞ or L^2, further difficulties arise. In this scenario b must not be in the range of A anymore, or the inverse of A may not be continuous.

Remark 5.2

Let H be an infinite dimensional Hilbert space. If a compact operator T on H is invertible, its inverse T^{-1} is unbounded.

Proof. Let e_n be an orthonormal sequence in H. Then $e_n \rightharpoonup 0$ and $Te_n \to 0$, but for all $n \in \mathbb{N}$

$$\|T^{-1}Te_n\| = \|e_n\| = 1,$$

cf. [Ped00]. $\qquad\qquad\square$

Jacques Hadamard formulated in 1902, what he called *un problème bien posé*, which is translated as a well-posed problem, cf. [Had02]. A problem is said to be well-posed, if it fulfills the following conditions for admissible data.

1. A solution exists.

2. The solution is unique.

3. The solution depends continuously on the data.

Depending on the problem, it has to be clarified, what is called admissible data, a solution and which topology defines continuity. This loose definition is frequently quoted in articles and books about inverse problems. The reason is that a violation of any of the above criteria leads to an ill-posed problem. The textbooks [EHN00] and [Rie03] are recommended as complementary literature for the theory of ill-posed inverse problems.

A famous example of ill-posedness in an everyday situation for mathematicians, and a verification of the ill-posedness of Volterra integral equations of the first kind is the differentiation of a smooth function, as it is described, e.g. by Bernd Hofmann in [Lue95]. In this example, the integral kernel k in (4.35) is the identity.

Example 1

Let $x \in \mathbb{C}^1([0,1])$, with $x(0) = 0$ and $y(t) = x'(t)$. For $0 \leq s \leq 1$, this problem can be written as a linear Volterra integral equation of the first kind,

$$x(s) = \int_0^s y(t)dt.$$

An observation $x^\delta \in \mathbb{C}^1([0,1])$ of the solution x may now be perturbed, such that for $\delta, n > 0$

$$x^\delta(t) = x(t) + \delta \sin(nt),$$

which satisfies

$$\|x^\delta - x\| = \delta.$$

The solution of the perturbed equation is then given by

$$y^\delta(t) = x'(t) + \delta n \sin(nt),$$

and thus,

$$\|y^\delta - y\| = \delta n.$$

Since n may be arbitrarily large, minimal errors in the observation yield misleading solutions.

Another example of an ill-posed problem can be found in [LRSs86]. There, Lavrent'ev et al. use the simple process of measuring a time-dependent physical field to illustrate ill-posedness and its consequences to the reader.

Example 2

Let the connection between a signal $\varphi(t)$, arriving at the input of some instrument, and the function $f(t)$, measured at the output given by the Volterra integral equation of the first kind

$$\int_0^t g(t - \tau)\varphi(\tau)d\tau = f(t). \tag{5.1}$$

Theoretically, the transition function g is obtained in the case where the input φ is equal to Dirac's delta distribution, which is not realizable in practice. Assuming g and f to be Schwartz functions, φ may be obtained by applying the Fourier transform and the convolution rule, see corollary 2.5. Then,

$$\varphi(t) = \frac{1}{2\pi} \int_{-\infty}^{\infty} e^{i\omega t} (\mathscr{F}\varphi)(\omega)d\omega = \frac{1}{2\pi} \int_{-\infty}^{\infty} e^{i\omega t} \frac{(\mathscr{F}f)(\omega)}{(\mathscr{F}g)(\omega)} d\omega. \tag{5.2}$$

The Fourier transformed of g tends to zero, as ω tends to zero. Thus, an arbitrarily small amount of noise in determining $(\mathscr{F}f)(\omega)$ for sufficiently large ω leads to large variations in the solution φ. In practice, the presence of white noise may cause such a problem.

The theory for linear operators is rather complete, as it is mentioned in [EHN00, Rie03] and [LRSs86]. Inverse problems related to vibrational analysis and structural dynamics may also be nonlinear and thus, a brief recall of the relevant concepts for both theories will be given in this section. An operator, denoted by F may be nonlinear, whereas A always denotes a linear operator.

In general, an operator equation

$$F(x) = y \tag{5.3}$$

is considered the forward problem, where $F : D(F) \subset X \to Y$ is a bounded operator between Hilbert spaces X and Y. Additionally, F is assumed to be weakly closed, but from now on, the presence of noise is always assumed. For a known, or estimated noise level $\delta > 0$ and $y \in R(F)$ consider y^δ, such that

$$\|y^\delta - y\| \leq \delta \tag{5.4}$$

in an appropriate norm. Solvability of equation (5.3) is always taken for granted, i.e. $x \in D(F)$ implies $F(x) = y \in R(F)$, and the inverse problem is to reconstruct $x \in X$ for a given element in Y such that (5.3) holds. Since the noisy observation $y^\delta \in Y$ is not necessarily an element of $R(F)$, an $x \in X$ that minimizes the defect

$$\|F(x) - y^\delta\| \tag{5.5}$$

may serve as an estimation. In general, this problem does not have a unique solution.

Definition 5.2
Let $F : X \to Y$ be a bounded operator between Hilbert spaces X and Y,

an arbitrary $y \in \mathrm{ran}(F) \oplus \mathrm{ran}(F)^{\perp}$ and $x^{*} \in X$. An $x^{\dagger} \in X$ is called x^{*}-*minimum-norm solution* of the inverse problem, if $F(x^{\dagger}) = y$ and

$$\|x^{\dagger} - x^{*}\| = \min\{\|x - x^{*}\| : F(x) = y\}.$$

If F is nonlinear, an x^{*}-minimum-norm solution does not necessarily exist, and existence does not assure uniqueness. The special case, where $x^{*} = 0$ will be just called *minimum-norm solution*.

5.1.1 Ill-posed problems

Since an ill-posed problem does not necessarily have a solution in the classical sense, the more general concept of a minimum-norm solution was introduced. Unfortunately, it may happen that such an approximation is not good enough in the presence of noise, cf. examples 1 and 2. For that matter, another idea is to approximate the original ill-posed problem by a family of similar, but well-posed problems. The following definition has been stated in various forms, e.g. in [EHN00, Rie03].

Definition 5.3
Let $A : X \to Y$ be a bounded linear operator between the Hilbert spaces X and Y. For $g \in R(A)$, let $f^{\dagger}$ be a minimum-norm solution of the inverse problem $Af = g$. For $\alpha_0 \in \mathbb{R}^{+}$ let $\{R_{\alpha}\}_{0 < \alpha < \alpha_0}$ be a family of continuous operators

$$R_{\alpha} : Y \to X.$$

The family $\{R_{\alpha}\}_{0 < \alpha < \alpha_0}$ is called a *regularization*, if there exists a *parameter choice rule* $\alpha = \alpha(\delta, g^{\delta})$ such that

$$\limsup_{\alpha \to 0}\{\|f^{\dagger} - R_{\alpha(\delta, g^{\delta})}(g^{\delta})\|_X : g^{\delta} \in Y, \|g - g^{\delta}\|_Y \leq \delta\} = 0.$$

A regularization and a parameter choice rule assure that regularized solutions converge in the norm towards the minimum-norm solution, if the noise level tends to zero. A prominent example of a regularization is the so called Tikhonov regularization.

For a given noisy measurement $g^\delta \in Y$, an initial guess $f_0 \in X$ and parameter $\alpha > 0$,

$$J_\alpha(f) := \|Af - g^\delta\|_Y^2 + \alpha\|f - f_0\|_X^2 \qquad (5.6)$$

is called *Tikhonov functional*. For any $\alpha > 0$, this functional is strictly convex and coercive, which ensures the existence of a global minimizer. This global minimizer, say f_α^δ, satisfies

$$f_\alpha^\delta = (A^*A - \alpha)^{-1}A^*g\delta, \qquad (5.7)$$

where A^* denotes the adjoint of A and the sole α is understood as α times the identity. The expression is obtained by setting the derivative of (5.6) zero, since the necessary and sufficient conditions coincide for this case.

The question of whether minimizing the Tikhonov functional is indeed a regularization of the ill-posed problem $Af = g^\delta$ was answered e.g. in [EHN00].

Theorem 5.3

Given noisy measurements g^δ, let f_α^δ be the minimizer of the Tikhonov functional (5.6) and let $g \in ran(A)$ with $\|g - g^\delta\| \leq \delta$. If $f^\dagger$ denotes the minimum-norm solution of $Af = g$ in the according norm, then the choice of $\alpha(\delta)$, such that

$$\lim_{\delta \to 0} \alpha(\delta) = 0 \quad and \quad \lim_{\delta \to 0} \frac{\delta^2}{\alpha(\delta)} = 0$$

yields

$$\lim_{\delta \to 0} \|f_\alpha^\delta - f^\dagger\|_X = 0.$$

Proof. See [EHN00]. ☐

An extension of Hadamard's characterization of a well-posed problem to nonlinear operators is given, e.g. in [Rie03], as follows.

Definition 5.4

Let X and Y be Banach spaces. The nonlinear operator equation (5.3) for X and Y is called *locally well-posed in* $f^+ \in D(F)$, if a $r > 0$ exists, such that for all sequence $\{f_k^r\}_{k \in \mathbb{N}} \subset B_r(f^+) \cap D(F)$

$$\lim_{k \to \infty} \|F(f_k^r) - F(f^+)\|_Y = 0 \implies \lim_{k \to \infty} \|f_k^r - f^+\|_X = 0. \qquad (5.8)$$

If an $f^+ \in D(F)$ exists, such that the operator equation (5.3) for X and Y is not locally well-posed in $f^+ \in D(F)$, it is said to be *locally ill-posed in* $f^+ \in D(F)$.

The theory of regularizing nonlinear ill-posed problems is not as well-arranged and complete as it is for the linear case. Nevertheless, the concept of Tikhonov regularization is transferable to nonlinear problems. At this point, [Rie03] is recommended as continuative literature.

5.2 Model updating

The system matrices, emerging from the FEM, depend on various material and geometrical parameters as it was shown in section 4.1. In order to fulfill many purposes in modern production facilities, different kinds of steel with varying combinations of alloy metals are employed. Due to the complexity of these machines, simplified geometric properties are often assumed, and since deconstructing these machines is often impossible, some components's dimensions have to be estimated.

A comprehensive survey of different methods and their history in the context of engineering was given by Mottershead and Friswell in 1993, cf. [MF93], and by Peter Verboven in his dissertation "Frequency-Domain System Identification for Modal Analysis", submitted in 2002 at the Vrije Universiteit Brussel in Belgium, cf. [Ver02]. In [Gla97], Gladwell supposed the mathematical aspect of this problem to be the construction of symmetric matrices with a band structure from given eigenvalues. The presented method therein is limited to so called *in-line systems*, i.e. finite element models where each element is restricted to having only 1 DOF at each end. Additional information on the problem of constructing symmetric matrices with a band structure from given eigenvalues and singular values can be found e.g. in [CG02, CdBY07].

In [TF08], Titurus and Friswell account for the ill-posedness of this problem and incorporate regularization. They briefly review the history of regularization methods in engineering applications and construct a theoretical setup for regularization. The article closes with some case studies. Today, this area does not seem to be completetly conquered, and an approach to regularization for finite element model updating is presented here. The results within this section resemble those in the diploma thesis "Verbesserung von Rotormodellen: Messdatenbasierte Modelladaption mit der Methodik Inverser Probleme" by Katthrin Arning, submitted at the University of Bremen in 2006, cf. [Arn06].

A closer look at two components of one element's stiffness matrix recalls the relevance of at least three different parameters. The elementwise torsion- and strain-stiffness are represented by

$$S_D = \frac{T}{l} \begin{bmatrix} 1 & -1 \\ -1 & 1 \end{bmatrix}, \quad S_T = \frac{B_x}{l} \begin{bmatrix} 1 & -1 \\ -1 & 1 \end{bmatrix},$$

depending on the torsional stiffness T, the strain stiffness B_x and the length l of the element, cf. subsection 4.1.2. Thus, it can be noted that system matrices of a structure, composed from different materials, depend on a parameter vector $p \in \mathbb{P} \subset \mathbb{R}^n$. In the case, where some a priori knowledge about the parameters in question is available, e.g. from literature or experience, let $p^0 \in \mathbb{P}$ contain these reference values or initial guesses and denote

$$\mathfrak{F}_s(p^0) := (M_0, D_0, S_0),$$

recalling (4.44).

The objective within this section is to adjust the reference parameters in a mathematical model such that simulations of the dynamical behavior match with measurements of the underlying existing structure. In practice, the parameters's range may be large. Steel may have a density of $d = 7.85e^{-3}\frac{g}{mm^3}$, while having an elastic modulus of $E = 210GPa \hat{=} 210.0e^9 \frac{g}{mms^2}$. Dealing with parameters within this range may cause cancellation of values close to the machine's precision, see [SB92]. In order to prevent this, relative changes in parameters will be considered from now on.

With $P_0 := \text{diag}(p^0)$, any parameter vector $p \in \mathbb{P}$ is obtained by

$$p = p^0 + P_0 x,$$

where $x \in \mathbb{R}^n$ is the offset to the reference value, and we have

$$p_i = p_i^0 + x_i \cdot p_i^0$$
$$\Longleftrightarrow \quad x_i = \frac{p_i - p_i^0}{p_i^0}.$$

Definition 5.5

Let $\mathfrak{F}_s : \mathbb{P} \to L(\mathbb{R}^s)^3$ be defined as in (4.44). Then the 3 matrices belonging to the triple

$$(\Delta M_i, \Delta D_i, \Delta S_i) := \mathfrak{F}_s(p_1, \ldots, p_{i-1}, 2p_i, p_{i+1}, \ldots, p_n) - \mathfrak{F}_s(p)$$

are called the *difference matrices of* $\mathfrak{F}_s(p)$.

If e.g. the j-th parameter has no impact on S, the according difference matrix ΔS_j will be the zero matrix.

The matrices M, D and S, resulting from an offset $x \in \mathbb{R}^n$ from the reference parameter p^0 can now be obtained by

$$M(x) := M_0 + \sum_{i=1}^{N} x_i \cdot \Delta M_i, \qquad (5.9)$$

$$D(x) := D_0 + \sum_{i=1}^{N} x_i \cdot \Delta D_i \qquad (5.10)$$

and

$$S(x) := S_0 + \sum_{i=1}^{N} x_i \cdot \Delta S_i. \qquad (5.11)$$

The definition above yields that a reduction of the i-th reference parameter
to e.g. 50%, gives the matrix

$$M_{i=0.5} = M_0 - 0.5 \cdot \Delta M_i$$

and analogously $D_{i=0.5}$ and $S_{i=0.5}$.

5.2.1 Harmonic loads

Consider an axially rotating spindel, described by the system

$$M(x)\ddot{u}(t) + D(x)\dot{u}(t) + S(x)u(t) = f(t). \qquad (5.12)$$

As an introduction to parameter identification, the force $f : \mathbb{R} \to \mathbb{R}^s$ on the
right hand side is restricted to only harmonic loads, i.e.

$$f(t) = f_0 \cdot e^{i\omega t},$$

where $f_0 \in \mathbb{R}^s$ contains the excentricity of an imbalances at given DOF. The
turning frequency ω is fix and arbitrary. These assumptions may appear
restrictive, but are fulfilled in many real life situations. Rotating drives or
spindles can be found in fans, chip removal machine tools and almost every
vehicle. The increasing of the turning frequency, for e.g. a starting engine of

a car or airplane is rather slow and can be assumed to be piecewise constant, see [DMM$^+$05]. It is clear that these requirements can be met easily in a laboratory environment.

Recalling section 4.2 and definition 3.4, the solution can be computed directly with the ansatz $u(t) = u_0 \cdot e^{i\omega t}$, $u_0 \in \mathbb{R}^s$. This results in

$$\left(-\omega^2 M(x) + i\omega D(x) + S(x)\right) u_0 = f_0$$
$$\Leftrightarrow u_0 = \left(-\omega^2 M(x) + i\omega D(x) + S(x)\right)^{-1} f_0, \tag{5.13}$$

and for $A_\omega(x)u_0 := \left(-\omega^2 M(x) + i\omega D(x) + S(x)\right) u_0$ a representation, like to that of $M(x), D(x)$ and $S(x)$ above can be given. It holds

$$A_\omega(x) = -\omega^2 \left(M_0 + \sum_{i=1}^{N} x_i \Delta M_i\right) + i\omega \left(D_0 + \sum_{i=1}^{N} x_i \Delta D_i\right) + \left(S_0 + \sum_{i=1}^{N} x_i \Delta S_i\right)$$

$$= -\omega^2 M_0 + i\omega D_0 + S_0 + \sum_{i=1}^{N} x_i \left(-\omega^2 \Delta M_i + i\omega \Delta D_i + \Delta S_i\right)$$

$$=: A_{\omega,0} + \sum_{i=1}^{N} x_i \Delta A_{\omega,i}. \tag{5.14}$$

The operator A_ω is a key element in the mathematical formulation of the model updating problem. It leads to the following definition.

Definition 5.6

Let $M(x), D(x)$ and $S(x)$ be defined as in (5.9), (5.10) and (5.11), $f_0 \in \mathbb{R}^s$ and $\omega \in \mathbb{R}^+$. Then, the operator

$$\mathscr{A}_\omega : \mathbb{R}^n \supset \mathbb{P}_\omega \longrightarrow \mathbb{R}^s,$$
$$\mathscr{A}_\omega(x) := A_\omega(x)^{-1} f_0$$

is called the *parameter-to-solution operator*.

Fixing the turning frequency ω is crucial for the analysis of the parameter-to-solution operator. It determines the domain of the operator, since $A_\omega(x)$

is not invertible if ω is an eigenfrequency of $M(x), D(x)$ and $S(x)$, cf. definition (4.5). As long as ω is not an eigenfrequency, the a parameter-to-solution operator is a composition of continuous operations and thus continuous. For given harmonic load f_0 and turning frequency ω it maps parameter offsets to the amplitudes u_0 of the vibrational responses in terms of (5.13). The occurrent inverse of A_ω makes it nonlinear, and its values tend to infinity when changing the parameter shifts an eigenfrequency near the turning frequency.

Lemma 5.4

The parameter-to-solution operator from definition 5.6 is totally differentiable, and its total derivative $\mathscr{A}'_\omega$, given by

$$\mathscr{A}'_\omega(x)h = -\sum_{i=1}^{n} h_i \cdot \left(A_\omega(x)^{-1}\Delta A_{\omega,i}A_\omega(x)^{-1}f_0\right)$$

is Lipschitz continuous.

Proof. It holds

$$A_\omega(x)^{-1}A_\omega(x) = \mathbb{1}$$
$$\Longleftrightarrow \quad \left(A_\omega(x)^{-1}A_\omega(x)\right)' = 0$$
$$\Longleftrightarrow \quad \left(A_\omega(x)^{-1}\right)'A_\omega(x) + A_\omega(x)^{-1}A_\omega(x)' = 0$$
$$\Longleftrightarrow \quad \left(A_\omega(x)^{-1}\right)'A_\omega(x) = -A_\omega(x)^{-1}A_\omega(x)'$$
$$\Longleftrightarrow \quad \left(A_\omega(x)^{-1}\right)' = -A_\omega(x)^{-1}A_\omega(x)'A_\omega(x)^{-1}.$$

Hence, with the representation of $A_\omega(x)$ given in (5.14), $A_\omega(x)'$ is given by

$$A'_\omega(x)h = A_\omega(x+h) - A_\omega(x) = \sum_{i=1}^{n} h_i \Delta A_{\omega,i},$$

and from Taylor's theorem the equality

$$A_\omega(x+h)^{-1} - A_\omega(x)^{-1} = \left(A_\omega(x)^{-1}\right)'h + \mathcal{O}(\|h\|^2)$$

is obtained. Thus, the directional derivatives of $\mathscr{A}_\omega$ in every $x \in \mathbb{P}_\omega$, are computed by

$$\lim_{t \to 0} \frac{\mathscr{A}_\omega(x + th) - \mathscr{A}_\omega(x)}{t} = \lim_{t \to 0} \frac{\left(A_\omega(x + th)^{-1} - A_\omega(x)^{-1}\right) f_0}{t}$$

$$= \left(\left(A_\omega(x)^{-1}\right)' h + \lim_{t \to 0} \frac{\mathcal{O}(\|th\|^2)}{t}\right) f_0$$

$$= -\sum_{i=1}^{n} h_i \left(A_\omega(x)^{-1} \Delta A_{\omega,i} A_\omega(x)^{-1}\right) f_0, \quad (5.15)$$

which is a composition of continuous operations for all i, and thus the total derivative of $\mathscr{A}_\omega$ is given by (5.15). The estimate

$$\|\mathscr{A}_\omega'(x) - \mathscr{A}_\omega'(y)\| = \sup_{h \in B} \|\mathscr{A}_\omega'(x)h - \mathscr{A}_\omega'(y)h\|_2$$

$$= \sup_{h \in B} \left\| -\sum_{i=1}^{n} h_i \left(\varphi_i(x) + \varphi_i(y)\right) f_0 \right\|_2$$

$$\leq \sup_{h \in B} \left\{ n \max_i \{|h_i| \cdot \|\varphi_i(x) - \varphi_i(y)\|_{\mathbb{R}^s \to \mathbb{R}^m} \} \|f_0\|_2 \right\}$$

$$= C \cdot \max_i \|\varphi_i(x) - \varphi_i(y)\|_{\mathbb{R}^s \to \mathbb{R}^m},$$

with an appropratiately chosen constant $C \in \mathbb{R}$, verifies the Lipschitz continuity of $\mathscr{A}_\omega'$ for any x and y in a bounded subset $B \subset \mathbb{P}_\omega$. $\qquad \square$

The proof of the existence and Lipschitz continuity of the total derivative of $\mathscr{A}_\omega$ is contained in [Arn06]. It is restated here, since the original source is not published. The restriction to a bounded subset $B \subset \mathbb{P}_\omega$ does not limit the applicability to real-life situations, since neither mass nor stiffness or damping can be infinte or zero in reality. Further, it is most likely in an application that measurements cannot be taken at every DOF. Therefore, a projection matrix $Q : \mathbb{R}^s \to \mathbb{R}^m$ is included that maps the simulation space $\mathbb{R}^s$ onto the measurement space $\mathbb{R}^m$. For matters of convenience, a possible inclusion of a projection Q and the mapping $Q\mathscr{A}_\omega : \mathbb{P}_\omega \to \mathbb{R}^m$ will also be denoted as $\mathscr{A}_\omega$. Nevertheless, the operator is a map between finite dimensional spaces and thus, it is locally well-posed in every point, see definition 5.4! The nonlinearity of the parameter-to-solution operator and its behavior around certain parameters may still result in an ill-condition that calls for regularization. For this reason, a Tikhonov functional will be defined.

Given noisy measurements u^δ, an $\alpha > 0$ and an initial guess $x^* \in \mathbb{P}_\omega \subset \mathbb{R}^n$, the according Tikhonov functional reads

$$J_\alpha(x) := \|\mathscr{A}_\omega(x) - u^\delta\|_{\mathbb{R}^m}^2 + \alpha \|x - x^*\|_{\mathbb{R}^n}. \tag{5.16}$$

A potential minimizer of J_α indicated necessary changes in the reference parameters to match the simulation and measured vibrational response as a reaction to a known monofrequent excitation. The question, whether such a minimizer exists shall be answered, next.

The functional (5.16) is bounded from below by zero, and thus a sequence $\{x_n\}_{n\in\mathbb{N}} \subset \mathbb{P}_\omega$ exists that satisfies

$$n \longrightarrow \infty \quad \Longrightarrow \quad J_\alpha(x_n) \searrow \inf\{J_\alpha(x) | x \in \mathbb{P}_\omega\}.$$

Hence, the sequence $\{J_\alpha(x_n)\}_{n\in\mathbb{N}}$ is monotonously decaying. Moreover, it holds

$$\|x_n\|_{\mathbb{R}^n} \leq \|x_n - x^*\|_{\mathbb{R}^n} + \|x^*\|_{\mathbb{R}^n}$$

$$\leq \sqrt{\frac{1}{\alpha} J_\alpha(x_n)} + \|x^*\|_{\mathbb{R}^n} \leq \sqrt{\frac{1}{\alpha} J_\alpha(x_1)} + \|x^*\|_{\mathbb{R}^n},$$

and

$$\|\mathscr{A}_\omega(x_n)\|_{\mathbb{R}^m} \leq \|u^\delta\|_{\mathbb{R}^m} + \sqrt{J_\alpha(x_n)} \leq \|u^\delta\|_{\mathbb{R}^m} + \sqrt{J_\alpha(x_1)}.$$

The theorem of Bolzano-Weierstraß now yields that $\{x_n\}_{n\in\mathbb{N}}$ has a convergent subsequence $\{x_{n_i}\}_{i\in\mathbb{N}}$, such that $\{\mathscr{A}_\omega(x_{n_i})\}_{i\in\mathbb{N}}$ converges, too. The respective limits of these sequences will be denoted by $\tilde{x} \in \mathbb{R}^n$ and $\tilde{u} \in \mathbb{R}^m$.

The continuity of $\mathscr{A}_\omega$ was indirectly verified in lemma (5.4). It yields that $\tilde{x} \in \mathbb{P}_\omega$ and $\mathscr{A}_\omega(\tilde{x}) = \tilde{u}$.

Finally, it holds

$$J_\alpha(\tilde{x}) = \lim_{i \to \infty} J_\alpha(x_{n_i}) = \inf\{J_\alpha(x) | x \in \mathbb{P}_\omega\},$$

and the existence of a minimizer of the Tikhonov functional (5.16) is verified.

For the minimization of the Tikhonov functional, the gradient method can be chosen, due to the parameter-to-solution operator being totally differentiable. The gradient of the Tikhonov functional is given by

$$\nabla J_\alpha(x) = 2\left(\mathscr{A}'_\omega(x)^*(\mathscr{A}_\omega(x) - u^\delta) + \alpha \cdot (x - x^*)\right). \tag{5.17}$$

The adjoint of $\mathscr{A}'_\omega(x)$ has been presented in [Arn06]. Here, the according lemma and its proof are cited.

Lemma 5.5

The adjoint of the derivative of the parameter-to-solution operator from lemma 5.4 is given by

$$\mathscr{A}'_\omega(x)^* y = \sum_{i=1}^{n} \langle \mathscr{A}'_\omega(x)e_i, y\rangle e_i. \tag{5.18}$$

Proof. It holds for $x, y, z \in \mathbb{R}^n$

$$\langle \mathscr{A}'_\omega(x)z, y\rangle = \sum_{i=1}^{n} \left(\mathscr{A}'_\omega(x)z\right)_i y_i$$

$$= \sum_{i=1}^{n} \left(-\sum_{j=1}^{n} z_j \left(A_\omega(x)^{-1}\Delta A_{\omega,j}A_\omega(x)^{-1}f_0\right)_j y_i\right)$$

$$= \sum_{j=1}^{n} z_j \left(-\sum_{i=1}^{n} \left(A_\omega(x)^{-1}\Delta A_{\omega,j}A_\omega(x)^{-1}f_0\right)_j y_i\right)$$

$$= \sum_{j=1}^{n} z_j \langle \mathscr{A}'_\omega(x)e_j, y\rangle = \left\langle z, \sum_{j=1}^{n} \langle \mathscr{A}'_\omega(x)e_j, y\rangle \right\rangle.$$

☐

This paves the way for a successful update of finite element model parameters. Unfortunately, no numerical results or simulations can be presented here.

5.3 Identification of load transmissions

In the preceding section, uncertainty about the correctness of the finite element model was the motivation. The mathematical methods, used therein belong to the field of parameter identification, and within a mathematical model of a mechanical system, the excitation can be considered a parameter, too. Nevertheless, the problem of identifying load transmissions from a measured vibrational responses is treated separately, here.

The identification of load transmissions may be very important in some situations and has various applications, e.g. narrowing the field of possible sources of unwanted excitations. The localization of imbalances in a nonlinear model of a gas turbine has been discussed by several researchers, cf. [DMM+05, RDM+06]. Furthermore, executing and evaluating an experimental modal analysis requests knowledge about the load transmission, e.g in order to compute the receptance, cf. definition 3.6. The problem of dealing with unknown load transmission brought up different approaches for output-only modal analysis, as it is addressed in [HdA99]. With the future goal of an instationary modal analysis in mind, a method for identifying arbitrary excitations of a mechanical structure from measurements of its vibrational response is valuable, cf. [Li13]. This section contains the main contribution to the fields of structural dynamics and vibrational analysis.

In order to focus on the identification of load transmissions, the dynamical behavior of the objective mechanical system is assumed to be modeled with sufficient accuracy by the system of ODE

$$M\ddot{u}(t) + D\dot{u}(t) + Su(t) = f(t), \qquad (5.19)$$

with real, symmetric and positive-definite matrices M, D and S of dimension s. Moreover, D is chosen to be a damping matrix, i.e. the matrix A, resulting from transforming (5.19) into a first order ODE system, is Hurwitz, cf. (4.39). As in previous formulations of this equation, the solution u of this ODE is a vibrational response to a load transmission f.

A question that might immediately come to mind is, why is recovering load transmissions from measured vibrational responses even an inverse problem at all? Regarding equation (5.19), computing the excitation f from measured dynamical behavior u^δ may appear to be a forward problem. But that the challenge of this section is an inverse problem becomes clearer after the transformation of (5.19) into a system of ODE of order one, as described in section 4.2. Vibrational responses are computed as solutions of the IVP

$$\dot{x}(t) = Ax(t) + F(t) \qquad (5.20)$$

with initial value $x(0) = 0$, and a constant system matrix A. As it is put nicely by Andreas Rieder in [Rie03], an inverse problem is understood as observing an effect and concluding its origin.

By accepting (5.20) as the model for simulations of the dynamical behavior of the mechanical system, described by the matrix A, the theorem of Caratheodory, see theorem 4.3, admits all measurable functions as possible load transmissions. Although there exists an endeavor to permit as many different kinds of load transmission as possible, the space of load transmissions is chosen to be $L^2(\mathbb{R}_0^+, \mathbb{R}^d)$ which is a subspace of all measurable functions. It will be called the *parameter space*. With load transmissions being represented by elements of $L^2(\mathbb{R}_0^+, \mathbb{R}^d)$, the solutions are elements in $AC_0(\mathbb{R}_0^+, \mathbb{R}^d) \subset L^2(\mathbb{R}_0^+, \mathbb{R}^d)$, due to theorem 4.3. The operator, defined in the following, will be used to examine how variations of the excitation influence the solution of (5.20).

Theorem 5.6

Let A be a real and non-singular matrix of rank d. If all eigenvalues of A have negative realpart, the operator, defined by

$$\mathscr{A}_A : L^2(\mathbb{R}_0^+, \mathbb{R}^d) \longrightarrow AC(\mathbb{R}_0^+, \mathbb{R}^d) \cap L^2(\mathbb{R}_0^+, \mathbb{R}^d),$$

$$(\mathscr{A}_A F)(t) := \int_0^t e^{(t-s)A} F(s)ds, \qquad (5.21)$$

is well-defined and Lipschitz-continuous. It is called the load-to-solution operator.

Proof. It holds

$$\|\mathscr{A}_A F - \mathscr{A}_A \tilde{F}\|_{L^2}^2 = \int_0^\infty \left\| (\mathscr{A}_A F)(t) - (\mathscr{A}_A \tilde{F})(t) \right\|_2^2 dt$$

$$= \int_0^\infty \left\| \int_0^t e^{(t-s)A} F(s)ds - \int_0^t e^{(t-s)A} \tilde{F}(s)ds \right\|_2^2 dt$$

$$= \int_0^\infty \left\| \int_0^t e^{(t-s)A} \left(F(s) - \tilde{F}(s) \right) ds \right\|_2^2 dt.$$

The product inside the norm is a convolution of $c_1(s) := e^{sA}$ and $c_2(s) := F(s) - \tilde{F}(s)$. Thus, Young's inequality can be applied to show that the norm is bounded from above, cf. theorem 2.6. It holds

$$\left\| \int e^{(\cdot - s)A} \left(F(s) - \tilde{F}(s) \right) ds \right\|_{L^2}$$

$$\leq \int \left\| e^{sA} \right\|_2 ds \int \|F(s) - \tilde{F}(s)\|_2^2 ds. \qquad (5.22)$$

The L^1-norm of e^{sA} is finite since all of the eigenvalues of A have negative realpart. The integral is computed component-wise and for every component it holds

$$\int_0^\infty |e^{\lambda s}| ds = \lim_{t \to \infty} \left| \frac{1}{\lambda} \left(e^{\lambda t} - 1 \right) \right|,$$

where λ is an eigenvalue of A. This eigenvalue is assumed to have negative realpart, and the exponential above may be written as

$$e^{x+iy} = e^{xt}e^{iyt} = e^{xt}\left(\cos(yt) + i\sin(yt)\right)$$

with $\lambda = x + iy$. Clearly, the second factor is bounded for every t and due to the negativity of x, the first factor tends to zero when t tends to infinity. Together with (5.22), this proves the claim

$$\|\mathscr{A}_A F - \mathscr{A}_A \tilde{F}\|_{L^2} \leq C \|F - \tilde{F}\|_{L^2}.$$

$\square$

The load-to-solution operator is a special case of the previously mentioned Volterra integral equations of the first kind. It is well documented that such an operator is compact, and its inversion is one of the most prominent ill-posed problems. A survey on different methods for regularizing Volterra integral equations of the first kind is given by Patricia Lamm in [Lam00]. The version at hand is special, because the kernel does not directly depend on both arguments s and t, but only on their difference. This fact was also used in the proof of theorem 5.21, where the operator was understood as a convolution. In this terminology, its inversion is called a deconvolution. Additionally, the input F may be understood as the control of a dynamical system, which paves the way to the mathematical field of optimal control. Yet another formulation will be regarded, ought to bring more clarity to some of the following arguments. Therefore, a definition given by Kress in [Kre99] is restated.

Definition 5.7
An integral kernel $k : (\mathbb{R} \times \mathbb{R}) \to \mathbb{R}^d$ is called *weakly singular* by definition, if k is well-defined and continuous for all $s, t \in \mathbb{R}, s \neq t$, and there exist positive constants C and $a \in (0, 1]$, such that

$$|s - t| > 0 \implies |k(s, t)| \leq C|s - t|^{-a}.$$

Incorporating the characteristic function

$$\chi_{[0,t]}(s) \longrightarrow \begin{cases} 1, & s \in [0,t] \\ 0, & \text{else} \end{cases} \tag{5.23}$$

for any $t > 0$, the load-to-solution can be reformulated as a *Fredholm integral equation*

$$(\mathscr{A}_A F)(t) = \int_0^t e^{(t-s)A} F(s)ds = \int_{\mathbb{R}} k(s,t)F(s)ds, \tag{5.24}$$

with the weakly singular integral kernel $k(s,t) := \chi_{[0,t]}(s)e^{(t-s)A}$. The key difference between Fredholm and Volterra integral equations is that the Fredholm integral limits are constant.

Either way, regularization is needed for inverting the load-to-solution operator, and standard Tikhonov regularization is applicable for this problem, cf. [Lam00]. Given noisy measurements x^δ the according Tikhonov functional is

$$J_\alpha(F) := \|\mathscr{A}_A F - x^\delta\|_{L^2}^2 + \alpha\|F\|_{L^2}^2, \tag{5.25}$$

and its minimum is given by

$$F_\alpha^\delta = (\mathscr{A}_A^* \mathscr{A}_A - \alpha)^{-1} \mathscr{A}_A^* x^\delta. \tag{5.26}$$

The regularization parameter can be chosen a-posteriori, i.e. after gathering the noisy data, such that

$$\left\|F_{\alpha(\delta,y^\delta)}^\delta - F^\dagger\right\| = \mathcal{O}\left(\delta^{\frac{2}{3}}\right).$$

The underlying theory is discussed thoroughly e.g. in [EHN00].

Computing the Hilbert space adjoint of the load-to-solution operator is the next step. A helpful lemma for this task, given in [Kre99] with less restrictive conditions, is reformulated here.

Lemma 5.7

For $d < \infty$ let $A : L^2(\mathbb{R}) \to L^2(\mathbb{R}^d)$ be defined by

$$(A\varphi)(\cdot) := \int_{\mathbb{R}} k(s, \cdot)\varphi(s)ds,$$

with a continuous or weakly singular integral kernel $k : (\mathbb{R} \times \mathbb{R}) \to \mathbb{R}^d$. Then the integral operator $B : L^2(\mathbb{R}) \to L^2(\mathbb{R}^d)$, defined by

$$(B\psi)(\cdot) := \int_{\mathbb{R}} k(\cdot, s)\psi(s)ds,$$

is the L^2-adjoint of A.

Since this is an exercise in [Kre99], the proof is given here.

Proof. For $d < \infty$, a weakly singular kernel $k : (\mathbb{R} \times \mathbb{R}) \to \mathbb{R}^d$ and $\varphi, \pi \in L^2(\mathbb{R})$, it holds for $a \in (0, 1]$

$$\int_{\mathbb{R}} \int_{\mathbb{R}} |k(s,t)\varphi(s)\psi(t)| ds dt = \left\| \psi \int_{\mathbb{R}} |k(s, \cdot)\varphi(s)| ds \right\|_{L^1}$$

$$\leq \|\psi\|_{L^2} \left\| \int_{\mathbb{R}} C|s - \cdot|^{-a}|\varphi(s)| ds \right\|_{L^2}$$

$$\leq C\|\psi\|_{L^2} \|\varphi\|_{L^2} \int_{\mathbb{R}} \int_{\mathbb{R}} |s - t|^{-2a} ds dt,$$

applying Hölder's inequality twice. Kress showed in [Kre99][1] that the double integral over the weakly singular kernel exists. Hence, Fubini's theorem can be applied to prove the lemma

$$\langle A\varphi, \psi \rangle_{L^2} = \int (A\varphi)(t)\psi(t)dt = \int \left(\int k(s,t)\varphi(s)ds \right) \psi(t)dt$$

$$= \int \varphi(s) \left(\int k(s,t)\psi(t)dt \right) ds = \int \varphi(s)(B\psi)(s)ds = \langle \varphi, B\psi \rangle_{L^2}.$$

$\square$

---

[1] In a more general form within the proof of theorem 2.22 on p.24

This result and (5.24) give way to the *adjoint load-to-solution operator*

$$\mathscr{A}_A^* : AC(\mathbb{R}_0^+, \mathbb{R}^d) \cap L^2(\mathbb{R}_0^+, \mathbb{R}^d) \longrightarrow L^2(\mathbb{R}_0^+, \mathbb{R}^d),$$

$$(\mathscr{A}_A^* G)(t) = \int_0^T k(t,s)G(s)ds = \int_t^T e^{(s-t)A}G(s)ds. \qquad (5.27)$$

5.3.1 Collocation method for Fredholm integral equations of the first kind

Nowadays, measurements are almost exclusively digital. Digital measurement processes are cheap and versatile, but they capture the reality only on predefined collocation points. In the context of structural dynamics, the distance of these collocation points is defined by the sampling rate, as explained in section 3.3.1. The purpose of this subsection is to give insight on how discrete vibrational response data can be used to solve the inverse problem of load transmission identification. The statements and conclusions within this subsection rely on the article "Regularized collocation method for Fredholm integral equations of the first kind" by M. Thamban Nair and Sergei V. Pereverzev, see [NP07]. A general overview of how to treat ill-posed problems numerically is given in [Rie03] within a chapter on projection methods. Projection methods themselves feature regularizing properties, cf. [EHN00, Rie03], but this is explained any further, here.

In the case where the available data is discrete, the original infinite dimensional problem is regarded on finite dimensional subspaces of the domain and range of the forward operator. In [NP07], the general Fredholm integral equation

$$(KF)(t) = \int_0^1 k(s,t)F(s)ds = x(t) \qquad (5.28)$$

for continuous, non-degenerate kernel k and $t \in [0,1]$ is considered to be the forward problem. The connection between equations like (5.28) and the problem of identifying load transmissions was revealed earlier in this section, cf. equation (5.24). The upper limit of the integral may represent a normed period of measurement. The idea in [NP07] is to approximate the integral (5.28) with a collocation scheme.

Collocation schemes, such as quadrature rules are in many cases the first choice for solving integral equations numerically, cf. [SB92, Ise98]. With an increasing sequence of collocation points $\tau_{i_i}^n \subset [0,1]$ and a sampling operator $\mathcal{T}_n : C([0,1]) \to \mathbb{R}^n$ such that for all $f \in C([0,1])$

$$\mathcal{T}_n f = (f(\tau_1)), f(\tau_2), \ldots, f(\tau_n)),$$

an operator $K_n = \mathcal{T}_n K : L^2(0,1) \to \mathbb{R}^n$ is defined. The ill-posedness of inverting (5.28) is reflected as an ill condition of the algebraic system $K_n F = \mathcal{T}_n x$, see [NP07]. Given noisy collocation data $x_n^\delta \in \mathbb{R}^n$, a minimizer of the Tikhonov functional according to K_n is obtained as a fixed point of

$$(K_n^* K_n + \alpha) F = K_n^* x_n^\delta,$$

for any $\alpha > 0$. It is shown in [NP07], how to compute the solution $F_{\alpha,n}^\delta \in L^2(0,1)$ of the equation above. As a main result, Nair and Pereverzev proof that the norm

$$\|F^\dagger - F_{\alpha,n}^\delta\|_{L^2} \tag{5.29}$$

is bounded. A choice for the regularization parameter α is suggested such that this bound depends on the assumed regularity of the solution $F^\dagger$, the chosen collocation scheme and the noise level, see theorem 1 in [NP07].

5.3.2 Discrete data and discrete solutions

In the previous subsection, the effect of discrete data on the solution of the inverse problem was shown. Additionally, an error estimate for the regularized solution was given. However, the load-to-solution operator originates from solving a system of ODE, and as described in section 4.2, solving ODE numerically may also be done by EOSM for reduced computational effort.

Hence, in this final subsection in the course of identifying load transmissions, the discretization is taken one step further with a more hands-on approach.

Assuming $\eta_0 := x_0, t_0 := 0$ and $k = 1, 2, \ldots T$, the IVP (4.37) can be solved numerically with an explicit one-step method (EOSM) Φ,

$$t_k := t_0 + kh, \quad \eta_{k+1} := \eta_k + h\Phi(A, f, \eta_k, t_k, h).$$

A general introduction to EOSM has been given in section 4.2.1. It is worth noting that the significance of a regularized solution of an ill-posed inverse problem depends on the quality of the forward model. EOSM *approximate* an analytical solution of an ODE. In the case discussed here, this ODE *approximates* the dynamical behavior of a mechanical system. Therefore, it has to be assumed that the mechanical system is not only modeled adequately by the matrices M, D and S, but also that the objective's dynamical behavior is described with satisfying accuracy by the solutions of (5.20), obtained with an EOSM.

The accuracy of the model and solutions, obtained by EOSM were discussed in the previous chapter. As mentioned there, the accuracy of every iterative method for solving differential equations depends on the step size, denoted by h in the formulation above. Smaller step sizes yield higher precision of the forward model, but regarding the inverse problem, the reciprocal of the sampling frequency of the measurement somehow gives a fixed step size.

In the case of Euler's method, where $\Phi(A, f, \eta_k, t_k, h) = A\eta_k + f(t_k)$, solutions can also be represented by a matrix-vector product. Therefore, given discrete load transmissions

$$f(t_1, \ldots, t_T) = \begin{bmatrix} F_{1,1} & \cdots & F_{T,1} \\ \vdots & \ddots & \vdots \\ F_{1,d} & \cdots & F_{T,d} \end{bmatrix} \in \mathbb{R}^{d \times T}$$

are ordered such that

$$
F := \begin{bmatrix} \begin{bmatrix} F_{1,1} \\ \vdots \\ F_{1,d} \end{bmatrix} \\ \vdots \\ \begin{bmatrix} F_{T,1} \\ \vdots \\ F_{T,d} \end{bmatrix} \end{bmatrix} .
\tag{5.30}
$$

Then, with the matrix

$$
E_T := h \cdot \begin{bmatrix} (\mathbb{1}_d + hA) & \mathbb{1}_d & & \\ \vdots & \vdots & \ddots & \\ (\mathbb{1}_d + hA)^T & (\mathbb{1}_d + hA)^{T-1} & \cdots & \mathbb{1}_d \end{bmatrix} ,
\tag{5.31}
$$

an approximate solution X of (5.20) is obtained by $X = E_T F$, where the ordering of X corresponds to that of F, i.e.

$$
X = \begin{bmatrix} \begin{bmatrix} x_{1,1} \\ \vdots \\ x_{1,d} \end{bmatrix} \\ \vdots \\ \begin{bmatrix} x_{T,1} \\ \vdots \\ x_{T,d} \end{bmatrix} \end{bmatrix} .
$$

The dimension of the vectors F and X is the product of twice the dimension of the model times the number of sample points. An equivalent representation can be found for other EOSM, but this is not pursued, here.

The ill-posedness of identifying load transmissions in the analytical formulation was shown in the beginning of section 5.3. In this subsection, the forward operator is a finite dimensional matrix and thus, the inverse problem is not ill-posed. Nevertheless, in order to obtain an inverse operator of E_T that produces significant results, regularization has to be applied. The ill-posedness of identifying load transmission using the model with an infinite dimensional domain, see theorem 5.6, is reflected as a severe ill condition of the finite dimensional matrix E_T. The reflection of the ill-posedness as an ill condition of the finite dimensional problem is illustrated with an example.

Example 3

Let a mechanical system with $s = d/2 \in \mathbb{N}$ DOF be modeled by $A \in \mathbb{R}^{d \times d}$, as seen in (4.30), and let $\eta \in \mathbb{R}^{2m \times T}$ be a discrete measurement of a vibrational response to an unknown load transmission at $m \leq s$ spots of the mechanical system. A lack of accessibility of the structure may provide measurements of displacements in less DOF than the mathematical model simulates.

For now, suppose $m = s$. Otherwise, a projection $P : \mathbb{R}^s \to \mathbb{R}^m$ has to be included. Applying Euler's method conversely, it holds for $k = 1, \ldots, T-1$

$$F_k^E = \frac{\eta_{k+1} - \eta_k}{h} - A\eta_k, \qquad (5.32)$$

and F_T^E may be set to zero. With unperturbed measurements $\eta_1, \cdots, \eta_T$ at hand, the load transmission F can be recovered qualitatively with this method. The graphs in figure 5.1 show the result of a forward application of Euler's method with a load transmission $F^\dagger$ as shown in the left of figure 5.2. The result of applying Euler's method conversely, i.e. equation (5.32), is shown in the right of figure 5.2. The difference between the original load transmission and its recovery, visible in figure 5.2, comes from the approximative character of Euler's method. Figure 5.1 shows the vibrational response at the respective points of the model A, where the forces were applied, cf. figure 5.2.

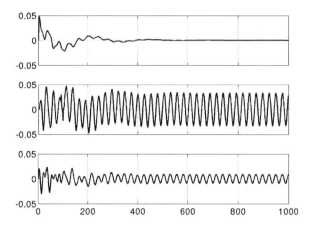

Figure 5.1: Vibrational response simulated by Euler's method

The presence of noise within the measurement leads to large variations in the recovery of the force. Figure 5.3 displays simulated measurements with added 0.5% Gaussian white noise and the resulting recovery of the force, obtained from applying Euler's method conversely.

The chosen regularization for this inverse problem again is to minimize an according Tikhonov functional, and the theoretical setting is explained next.

Assuming Euler's method to be the forward operator, discrete load transmission data $F \in \mathbb{R}^{d \times T}$ is mapped to samples of the resulting vibrational response, denoted by $X \in \mathbb{R}^{d \times T}$. The re-ordering, described in (5.30), is applied such that the load transmission and response are vectors in $\Theta_d := \mathbb{R}^{dT}$. The Tikhonov functional reads

$$J_\alpha(F) := \|E_T F - X^\delta\|^2_{\Theta_d} + \alpha \|F\|^2_{\Theta_d}. \tag{5.33}$$

The interpretation of the given data $X^\delta \in \mathbb{R}^{m \times T}$ is as follows. For i :

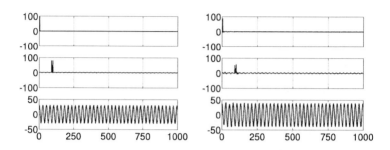

Figure 5.2: Original load transmission (left) and its recovery (right)

$1, \ldots, T$, the i-th column X_i^δ is a perturbed sample of an unknown, absolutely continuous function $x \in AC(\mathbb{R}_0^+, \mathbb{R}^d) \cap L^2(\mathbb{R}_0^+, \mathbb{R}^d)$ at given collocation points $\tau_i, \ldots, \tau_T$ and $m \leq d$ measurement points. Additionally, including a projection matrix $P : \mathbb{R}^d \to \mathbb{R}^m$, the difference $\|X_i^\delta - Px(\tau_i)\|$ is supposed to be smaller than the noise level δ in the according Euclidean norm. The regularized solution vector $F_{\alpha,T}^{\delta,E} \in \Theta_d$,

$$F_{\alpha,T}^{\delta,E} = (E_T^* E_T + \alpha)^{-1} E_T^* X^\delta, \tag{5.34}$$

represents samples of an unknown function $F^\dagger \in L^2(\mathbb{R}_0^+, \mathbb{R}^d)$ at $\tau_i, \ldots, \tau_T$, which satisfies

$$\mathscr{A} F^\dagger = x. \tag{5.35}$$

In (5.34), E_T^* denotes the transposed of E_T.

5.4 Input shaping

In the previous section, the general attempt to recover load transmissions from given vibrational response data was discussed. A special case of determining a certain load transmission is when the dynamical response is ought

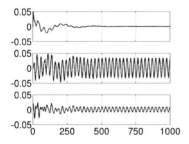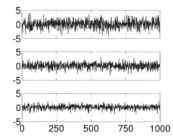

Figure 5.3: Noisy measurement (left) and recovery of the load transmission divided by 10^4 (right)

to have no vibrations at a certain frequency. This task has occupied scientists in the past and still is an active field of research, known as input shaping or impulse modeling.

A timeless motivation for preventing mechanical structures from vibrating is the ever decreasing size of electronic devices and the resulting necessity of higher precisions, [WB06]. On the other hand, production speed is increased, which results in increasingly irregular and quickly changing load transmissions in the production facilities. These kind of processes lead to wide-band excitations of the systems, where the motion of the machines themselves induce vibrations. The faster the machine moves and the quicker changes in motion are, the more energy is induced into the system and the larger resulting vibrations may become.

Regarding impulse-type stimulations, a series of N impulses may be written as

$$f(t) = \sum_{n=0}^{N} f_n e^{-\left(\frac{(t-t_n)}{\sigma_n}\right)^2}, \qquad (5.36)$$

with real, positive numbers f_n and t_n, representing force and time of impact. The parameter σ alters the width of the curve and should be chosen to adjust the energy of the impulse. The frequencies, excited by this kind of input can be seen through its amplitude spectrum, i.e. the modulus of its Fourier transform, which reads

$$\mathscr{F}f(\omega) := \frac{1}{2}\sigma e^{-\left(\frac{1}{2}\sigma\omega\right)^2}\left(f_0 + \sum_{n=1}^{N} f_n e^{-it_n\omega}\right). \tag{5.37}$$

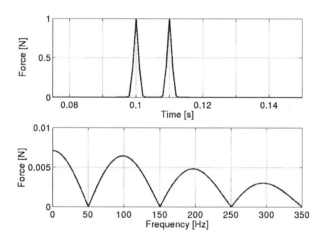

Figure 5.4: Two impulses and modulus of the corresponding frequency spectrum

The fact that this function attains zero, cf. figure 5.4, may lead to the idea of choosing the time-delays $t_0 - t_n, n \neq 0$ such that certain frequencies are not excited. For a fixed frequency ω_0, the series of delays can be computed via

$$\sum_{n=1}^{N} e^{-it_n\omega_0} = -f_0. \tag{5.38}$$

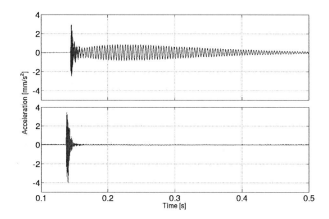

Figure 5.5: Vibrational response to a single and double impulse

The method of input shaping has been tested theoretically with FEM models and practically at the laboratory of the WZM. Figure 5.5 shows two measured dynamical reactions of the same structure. The blue line in figure 5.5 shows the measured vibrational response to a single impulse. The major resonance frequency ω_0 of the structure was determined from this measurement with the peak amplitude method. According to (5.38), a time-delay for a double impulse was computed that would prevent the structure from vibrating at ω_0 Hz. The red line in figure 5.5 shows the measured vibrational response to this double impulse. Both excitations were realized with an electromagnetic shaker.

The historic origin of this method was in 1988, when Neil C. Singer attempted in his dissertation [Sin88] to improve the performance of machines that exhibit flexibility by modifying their control. His objective was the robot arm of the U.S. Space Shuttle, which should not vibrate, once it stopped at its default position. The dynamical behavior until the endpoint of its path did not matter. For a given frequency ω_0, he derives an impulse series, such that a convolution of the original control sequence and the impulse series results in significant vibration reduction at ω_0.

His investigation started from the fact that vibrational responses of an uncoupled, linear mechanical system of any order can be specified as a cascaded set of second-order poles with the decaying sinusoidal response

$$u(t) = f_0 \left(\frac{\omega_0}{\sqrt{1 - c^2}} e^{-c\omega_0(t - t_0)} \right) \sin(\omega_0 \sqrt{1 - c^2}(t - t_0)),$$

where f_0 is the amplitude of the impulse, ω_0 is the natural frequency of the system, c is the damping ratio and t_0 is the instant of the impulse. The impulse response of two impulses may be superposed so that after the second impulse has ended, no residual vibrations of a certain frequency remain. The time-delays between two impulses, determined with his method, coincide with those, obtained from the Fourier transform approach (5.38).

The method of input shaping was further developed, to reduce residual vibrations in a wider range of frequencies. In subsequent publications about input shaping the focus is on how to improve the robustness of the method against errors in the estimates of damping and natural frequency [SS90, SSS94, MRB09], how input shaping improves trajectory following [SSS94] and how to create input shapers for multiple modes [MRB09, SCS97]. With this method, arbitrary work processes can be optimized. A convolution of any load transmission with a carefully chosen impulse series results in a load transmission, which causes minimal residual vibration after the work process has ended and less vibration during the process itself.

6 Conclusion

Most methods, used in applied mathematics today have their origin in the engineering sciences or physics. Hairer et al. remark in [HrW87] that the equation of motion, introduced in chapter 3 may be the first differential equation to ever concern scientist. The spatial discretization of mechanical systems may be the first application of the finite element method, as it is supposed by Zienkiewicz and Taylor in [ZT94]. Finally, the occurring inverse problem from section 5.3 is the Fredholm integral equation of the first kind, which is probably the first ill-posed inverse problem to be regularized.

Although the connection between mathematics and engineering are strong, the understanding between these groups of scientists is not always easy. This thesis contributes to the dialog between mathematicians and engineers in the field of structural dynamics. It is addressed to mathematicians as it provides a foundation of relevant physics that motivates mathematical research in structural dynamics. Moreover, the concepts taken from engineering literature are reformulated and conditions for mathematical validity are added. Finally, an overview of inverse problems in structural dynamics in both fields is given. In the following paragraphs, the look goes first to the results of this thesis and to open problems afterward.

6.1 Summary

From the beginning of chapter 3 until the end of chapter 4, the process of modeling dynamic behavior of complex mechanical structures is presented.

Through laws of physics, an equation is deduced that admits characterizing the principle dynamical behavior of a mechanical system with 1 DOF through its eigenfrequency and damping ratio. Along the way, methods for estimating model parameters from measured vibrational responses were such as stiffness and damping shown, cf. lemma 3.1 and 3.2 and equation (3.19). The theoretical results were verified through real measurements in section 3.3.

In 4.1 the concept is extended to more complex structures and a general approach to developing mathematical models for beam-like structures is presented. The proposed method was exemplified for an ultra precise milling machine in section 4.1.4. In section 4.3 the FEM models were augmented with a term, proportional to the systems velocity. It was shown that such an augmentation models a systems damping if and only if for given mass and stiffness matrices M and S all eigenvalues of

$$A := \begin{bmatrix} 0 & \mathbb{1} \\ -M^{-1}S & -M^{-1}D \end{bmatrix}$$

have negative real part, after proposition 4.4 by applying a theorem from [Son98].

The theories, presented in chapter 5 rely on a given certainty about the mathematical model. In section 5.2, the geometry of the structure is assumed to be captured correctly within the mathematical model, and knowledge about the load transmission is required. Under these conditions it is shown that material parameters of the model can be identified with Tikhonov regularization. These results rely on those made by Arning in [Arn06], while necessary additions regarding the domain of the forward operator were made. Further it has to be mentioned that in this case, the FEM can be seen as a *selective singular value decompostion*, and it turns the inversion into a well-posed, but ill-conditioned problem.

The identification of arbitrary load transmissions, section 5.3, marks the main theoretical contribution of this thesis to the field of structural dynamics.

The well-posedness, Lipschitz continuity and Hilbert space adjoint of the analytical forward operator were shown, see theorem 5.6 and lemma 5.7. This is the key to further approaches to the problem of identifying arbitrary load transmissions. The fact that most ODE are numerically solved by EOSM was the motivation for section 5.3.2. Here, an EOSM is assumed to suffice as a forward operator. This puts the problem of identifying load transmission on a completely discrete setup. The severe ill condition of the resulting equation was exemplified in 3.

In the final section, an introduction to input shaping is given. The presented verification via Fourier transform is not mentioned in any of the listed sources and seems to be new. Moreover, though 20 years later, the method of input shaping was obtained independently through this approach and without knowledge of the original sources.

6.2 Prospect

It worth noting that even in the 1 DOF case there are still open questions, such as the incorrectness of the assumed model, cf. figure 3.3. In order to get a first idea of what is going on, model updating under the assumption of dynamic parameters could be applied. It can be suggested from the measurements that the system in rest is stiffer, hence more direct responding to the load. Additionally, although the introduced half-power bandwidth method seems to be applied and analyzed frequently, none of the listed sources contains a mathematical verification. This might be a minor problem.

Regarding the proposed FEM models and their development, it has to be mentioned that none of the listed sources contained a thorough mathematical motivation of PDE (4.9). It is stated in [HK04] that the PDE can be deduced from the POVD (4.12) and vice versa. In [CH93] and [KW99] a similar PDE is obtained as a necessary condition for minimizing a mechanical system's potential energy, but only for simpler systems. Assuming

Hamilton's principle to be true, cf. [CH93], it has to be checked whether the PDE can be deduced as a necessary condition for a stationary point of an according energy functional.

In the proposed method it is assumed that the objective structure is beam-like, without giving a geometrical constraint on what beam-like means. Here, a thorough mathematical definition is needed to ensure a certain quality of the final model. The result for verifying whether an augmentation of the finite element model qualifies as a damping may be evolved into a more practical criterion, cf. 4.4. Numerical simulations indicate that Rayleigh damping, cf. 4.43, is a sufficient damping model in many cases, this has not been verified analytically and in general, yet.

The largest prospect in chapter 5 is an application to real data, though this has been done in case of input shaping, cf. 5.4. A fact that has been neglected in the sections 5.2 and 5.3 is that the measured dynamical behavior is not necessarily the systems displacement, but its velocity or acceleration. Thus, working on real data may demand lots of pre-processing.

Theoretically, 3 open problems seem to be the most interesting. First, the limitation of harmonic loads in section 5.2 has to be overcome. In general, the proposed method may work for any load transmission in L^2. Second, it may be analyzed whether the proposed method for load transmission identification is able to obtain the results from input shaping, given an according vibrational response. The third interesting problem is an analysis of the solutions, obtained from the inverse EOSM in the fashion of Nair and Pereverzev, cf. subsection 5.3.1.

Bibliography

[Ama90]   Herbert Amann, *Ordinary differential equations*, Walter de Gruyter Berlin & Co., Berlin - New York, 1990.

[AMF98]   Hamid Ahmadian, John E. Mottershead, and Michael I. Friswell, *Regularisation methods for finite element model updating*, Mechanical Systems and Signal Processing **12** (1998), no. 1, 47–64.

[Arn06]   Katthrin Arning, *Verbesserung von Rotormodellen: Messdatenbasierte Modelladaption mit der Methodik Inverser Probleme*, Master's thesis, Zentrum Für Technomathematik, Universität Bremen, Bremen, Germany, 2006.

[BG63]   Richard E. D. Bishop and Graham M. L. Gladwell, *An investigation into the theory of resonance testing*, Philosophical transactions of the Royal Society, London **255** (1963), 241–280.

[BK09]   Hans D. Baehr and Stephan Kabelac, *Thermodynamik*, Springer Berlin, 2009.

[Bos08]   Siegfried Bosch, *Lineare Algebra*, Springer Berlin, 2008.

[BS82]   Ivo Babuska and Barna Szabo, *On the rates of convergence of the finite element method*, International Journal for Numerical Methods in Engineering **18** (1982), no. November 1980, 323–341.

[Cau60]   Thomas K. Caughey, *Classical normal modes in damped linear dynamic systems*, J. appl. Mech. **27** (1960), no. 2, 269–271.

[CdBY07]   Moody T. Chu, Nicoletta del Buono, and Bo Yu, *Structured quadratic inverse eigenvalue problem, I. Serially linked systems*, SIAM Journal on Scientific Computing **29** (2007), no. 6, 2668–2685.

[CG02]   Moody T. Chu and Gene H. Golub, *Structured inverse eigenvalue problems*, Acta Numerica (2002), 1–71.

[CH93]   Richard Courant and David Hilbert, *Methoden der mathematischen Physik*, Springer Berlin, 1993.

[DMM⁺05] Volker Dicken, Ingo Menz, Peter Maaß, Jenny Niebsch, and Ronny Ramlau, *Nonlinear inverse imbalance reconstruction in rotordynamics*, Inverse Problems in Science & Engineering **13** (2005), no. 5, 507–543.

[DMZ94]   Geoffrey M. Davis, Stéphan G. Mallat, and Zhifeng Zhang, *Adaptive time-frequency decompositions*, SPIE Journal of Optical Engineering **33** (1994), no. 7, 2183–2191.

[EHN00]   Heinz W. Engl, Martin Hanke, and Andreas Neubauer, *Regularization of inverse problems*, Kluwer Academic Publishers, 2000.

[Fis02]   Gerd Fischer, *Lineare Algebra*, Vieweg, 2002.

[FMA01]   Michael I. Friswell, John E. Mottershead, and Hamid Ahmadian, *Finite-element model updating using experimental test data: parametrization and regularization*, Philosophical Transactions of the Royal Society **359** (2001), 169–186.

[Fri08]   Michael I. Friswell, *Inverse problems in structural dynamics*, Second International Conference on Multidisciplinary Design Optimization and Applications, no. September, 2008, pp. 1–12.

[Ger01]   Christian Gerthsen, *Physik*, Springer Berlin, 2001.

[GHS03]   Dietmar Gross, Werner Hauger, and Walter Schnell, *Technische Mechanik 1*, Springer Berlin, 2003.

[GK89a]   Robert Gasch and Klaus Knothe, *Strukturdynamik Band 1*, Springer Berlin, 1989.

[GK89b]   ———, *Strukturdynamik Band 2*, Springer Berlin, 1989.

[Gla97]   Graham M. L. Gladwell, *Inverse vibration problems for finite-element models*, Inverse Problems **13** (1997), 311–322.

[Gra04]   Wodek K. Grawonski, *Advanced structural dynamics and active control of structures*, Springer New York, 2004.

[Gro01]   Karlheinz Groechenig, *Foundations of time-frequency analysis*, Birkhäuser Boston, 2001.

[Gun81]   Albert F. Gunns, *The first Tacoma Narrows Bridge: A brief history of Galloping Gertie*, The Pacific Northwest Quarterly **72** (1981), no. 4, 162–169.

[Had02]   Jacques Hadamard, *Sur les problèmes aux dérivés partielles et leur signification physique*, Princeton University Bulletin **13** (1902), no. 2, 49–52.

[HdA99]   Luc Hermans and Herman Van der Auweraer, *Modal testing and analysis under operational conditions: Industrial applications*, Mechanical Systems and Signal Processing **13** (1999), no. 2, 193–216.

[HK04]   Friedel Hartmann and Casimir Katz, *Structural analysis with finite elements*, Springer Berlin Heidelberg, 2004.

[HP05]   Diederich Hinrichsen and Anthony J. Pritchard, *Mathematical systems theory I*, Springer Berlin, 2005.

[HrW87]   Ernst Hairer, Syvert P. Nørsett, and Gerhard Wanner, *Solving ordinary differential equations I*, Springer Berlin, 1987.

[Hur95]   Adolf Hurwitz, *Ueber die Bedingungen, unter welchen eine Gleichung nur Wurzeln mit negativen reellen Theilen besitzt*, Mathematische Annalen **46** (1895), no. 2, 273–284.

[Ise98]   Arieh Iserles, *A first course in the numerical analysis of differential equations*, Cambridge University Press, 1998.

[Kre99]   Rainer Kress, *Linear integral equations*, Springer New York, 1999.

[Kut01]   Martin W. Kutta, *Beitrag zur näherungsweisen Integration totaler Differentialgleichungen*, Z. Math. Phys. **46** (1901), 435–453.

[KW99]   Klaus Knothe and Heribert Wessels, *Finite Elemente*, Springer Berlin Heidelberg, 1999.

[Lam00]   Patricia K. Lamm, *A survey of regularization methods for first-kind Volterra equations*, Surveys on Solution Methods for Inverse Problems, Springer, 2000, pp. 53–82.

[LG95]   Man Liu and D. G. Gorman, *Formulation of Rayleigh damping and its extensions*, Computers & Structures **57** (1995), no. 2, 277–285.

[Li13]   Linghan Li, *Instationary modal analysis for impulse-type stimulated structures*, Ph.D. thesis, Universität Bremen, 2013.

[LKSK11]   Linghan Li, Bastian Kanning, Christian Schenck, and Bernd Kuhfuß, *Comparing different approaches for model parameters identification in short time*, International Symposium on Signal Processing and Information Technology 2011 (Adel Elmaghraby and Dimitrios N. Serpanos, eds.), IEEE, 2011.

[LRSs86]   Mikhail M. Lavrent'ev, Vladimir G. Romanov, and Sergeĭ P. Shishat-skiĭ, *Ill-posed problems of mathematical physics and analysis*, Amercian Mathematical Society, 1986.

[Lue95]   Heinz Luebbig (ed.), *The inverse problem*, Akademie Verlag, 1995.

[Mee08]   Karl Meerbergen, *Fast frequency response computation for Rayleigh damping*, Int. J. Numer. Meth. Engng **73** (2008), 96–106.

[Mei95]   Gjerrit Meinsma, *Elementary proof of the Routh-Hurwitz test*, Systems & Control Letters **25** (1995), 237–242.

[MF93]   John E. Mottershead and Michael I. Friswell, *Model updating in structural dynamics: A survey*, Journal of Sound and Vibration **167** (1993), no. 2, 347–375.

[Moe08]   Elvira Moeller, *Handbuch Konstruktionswerkstoffe*, Carl Hanser München, 2008.

[MR04]   Prasenjit Mohanty and Daniel J. Rixen, *Operational modal analysis in the presence of harmonic excitation*, Journal of Sound and Vibration **270** (2004), 93–109.

[MR06]   John E. Mottershead and Yitshak M. Ram, *Inverse eigenvalue problems in vibration absorption: Passive modification and active control*, Mechanical Systems and Signal Processing **20** (2006), 5–44.

[MRB09] Giovanni Mimmi, Carlo Rottenbacher, and Giovanni Bonandrini, *Theoretical and experimental sensitivity analysis of extra insensitive input shapers applied to open loop control of flexible arm*, International Journal of Mechanics and Materials in Design **5** (2009), 61–77.

[MSSS10] Peter Maaß, Chen Sagiv, Nir Sochen, and Hans-Georg Stark, *Do uncertainty minimizers attain minimal uncertainty?*, Journal of Sound and Vibration **16** (2010), no. 3, 448–469.

[NP07] M. Thamban Nair and Sergei V. Pereverzev, *Regularized collocation method for Fredholm integral equations of the first kind*, Journal of Complexity **23** (2007), no. 4-6, 454–467.

[OR10] Bertha A. Olmos and Jose M. Roesset, *Evaluation of the half-power bandwidth method to estimate damping in systems without real modes*, Earthquake Engineering and Structural Dynamics **39** (2010), 1671–1686.

[Ped00] Michael Pedersen, *Functional analysis in applied mathematics and engineering*, Champman & Hall/CRC, 2000.

[PRK85] A. Paulraj, R. Roy, and T. Kailath, *Estimation of signal parameters via rotational invariance techniques - ESPRIT*, Nineteeth Asilomar Conference on Circuits, Systems and Computers, vol. 33, IEEE, 1985, pp. 83–89.

[RDM+06] Ronny Ramlau, Volker Dicken, Peter Maaß, Carsten Streller, and Adrian Rienäcker, *Inverse imbalance reconstruction in rotordynamics*, ZAMM **86** (2006), no. 5, 385–399.

[Rie03]   Andreas Rieder, *Keine Probleme mit inversen Problemen*, Vieweg, 2003.

[Run95]   Carl Runge, *Ueber die numerische Auflösung von Differential-gleichungen*, Math. Ann **46** (1895), no. 2, 167–178.

[SAM+05]   Steven H. Strogatz, Daniel M. Abrams, Allan McRobie, Bruno Eckhardt, and Edward Ott, *Crowd synchrony on the Millennium Bridge*, Nature **438** (2005), no. 3, 43–44.

[SB92]   Josef Stoer and Roland Bulirsch, *Introduction to numerical analysis*, Springer Berlin, 1992.

[SCS97]   William E. Singhose, Ethan A. Crain, and Warren P. Seering, *Convolved and simultaneous two-mode input shapers*, IEE Proceedings Control Theory and Applications **144** (1997), 515–520.

[SF09]   Minas D. Spiridonakos and Spilios D. Fassois, *Parametric identification of a time-varying structure based on vector vibration response measurements*, Mechanical Systems and Signal Processing **23** (2009), 2029–2048.

[SGH02]   Walter Schnell, Dietmar Gross, and Werner Hauger, *Technische Mechanik 2*, Springer Berlin, 2002.

[She07]   Thomas R. Shearer, *The designer's guide to tungsten carbide*, Tech. report, General Carbide, August 2007.

[Sim81a]   Stepan S. Simonian, *Inverse problems in structural dynamics-I. Theory*, International Journal for Numerical Methods in Engineering **17** (1981), 357–365.

[Sim81b]   _____, *Inverse problems in structural dynamics-II. Applications*, International Journal for Numerical Methods in Engineering **17** (1981), 367–386.

[Sin88]   Neil C. Singer, *Residual vibration reduction in computer controlled machines*, Ph.D. thesis, Department of Mechanical Engineering, MIT, 1988.

[Son98]   Eduardo D. Sontag, *Mathematical control theory: Deterministic finite dimensional systems*, Springer New York, 1998.

[SS90]   Neil C. Singer and Warren P. Seering, *Preshaping command inputs to reduce system vibration*, ASME Journal of Dynamical Systems Measurement and Control **112** (1990), no. 1, 76–82.

[SSS94]   William E. Singhose, Warren P. Seering, and Neil C. Singer, *Residual vibration reduction using vector diagrams to generate shaped inputs*, ASME Journal of Mechanical Design **116** (1994), 654–659.

[TF08]   Branislav Titurus and Michael I. Friswell, *Regularization in model updating*, International Journal for Numerical Methods in Engineering **75** (2008), 440–478.

[TM01]   Françoise Tisseur and Karl Meerbergen, *The quadratic eigenvalue problem*, SIAM Rev. **43** (2001), no. 2, 235–286.

[Ver02]   Peter Verboven, *Frequency-domain system identification for modal analysis*, Ph.D. thesis, Vrije Universiteit Brussel, 2002.

[Wal00]   Wolfgang Walter, *Gewöhnlich Differentialgleichungen*, Springer Berlin, 2000.

[WB06]   Manfred Weck and Christian Brecher, *Werkzeugmaschinen - Messtechnische Untersuchung und Beurteilung, dynamische Stabilität*, Springer Berlin, 2006.

[Wer05]   Dirk Werner, *Funktionalanalysis*, Springer Berlin, 2005.

[WJ87]   William Weaver and Paul R. Johnston, *Structural dynamics by finite elements*, Prentice Hall, 1987.

[YB01]   Warren C. Young and Richard G. Budynas, *Roark's formulas for stress & strain*, McGraw-Hill Professional, 2001.

[YL05]   Jann N. Yang and Silian Lin, *Identification of parametric variations of structures based on least squares estimation and adaptive tracking technique*, Journal of Engineering Mechanics **131** (2005), no. 3, 290–298.

[Zha90]   Weijan Zhang, *Spectral density of the nonlinear damping model: Single dof case*, IEEE Transactions on Automatic Control **35** (1990), no. 12, 1320–1329.

[Zha94]   _____ , *Spectral density of a nonlinear damping model: Multi-dof case*, IEEE Transactions on Automatic Control **39** (1994), no. 2, 406–410.

[ZT94]   Olgierd C. Zienkiewicz and Robert L. Taylor, *The finite element method*, McGraw-Hill, 1994.